T0280523

Quantum Electrodynamics

This book provides an accessible introduction to quantum electrodynamics. Based on lectures on quantum electrodynamics given by the highly original and distinguished physicist V. N. Gribov, the aim of the book is to present the theory of quantum electrodynamics in the shortest and clearest way for applied use. A distinctive feature of Gribov's approach is the systematic use of the Green function method which allows a straightforward generalization to the cases of strong and weak interactions. The book starts with an introduction that uses the basics of quantum mechanics to introduce the reader gently into the world of propagation functions and particle interactions. The following chapter then focuses on spin $\frac{1}{2}$ particles. The text goes on to discuss symmetries, the CPT theorem, causality, and unitarity followed by a detailed presentation of renormalization theory. A final chapter looks at difficulties with the theory and possible routes to their resolution.

VLADIMIR NAUMOVICH GRIBOV received his Ph.D. in theoretical physics in 1957 from the Physico-Technical Institute in Leningrad where he had worked since 1954. From 1962 to 1980 he was the head of the Theory Division of the particle physics department of that institute, which in 1971 had become the Leningrad Institute for Nuclear Physics. In 1980 he moved to Moscow where he became the head of the particle physics section of the Landau Institute for Theoretical Physics. From 1981 he regularly visited the Research Institute for Particle and Nuclear Physics in Budapest where he was a scientific adviser until his death in 1997. Vladimir Gribov was one of the leading theoretical physicists of his time, who made seminal contributions to many fields, including quantum electrodynamics, neutrino physics, non-Abelian field theory, and especially to the physics of hadron interactions at high energies.

JULIA NYIRI received her Ph.D. in 1970 from JINR, Dubna. She first worked as a researcher and then from 1970 as senior scientist at the Central Research Institute for Physics of the Hungarian Academy of Sciences. From 1981 on, she has been senior scientist in the Research Institute for Particle and Nuclear Physics in Budapest. Her main fields of interest are: group theoretical problems, the quantum mechanical three-body problem, the quark structure of hadrons, and phenomenology of soft processes.

CAMBRIDGE MONOGRAPHS ON PARTICLE PHYSICS, NUCLEAR PHYSICS AND COSMOLOGY

13

General Editors: T. Ericson, P. V. Landshoff

QUANTUM ELECTRODYNAMICS

Gribov Lectures on Theoretical Physics

V. N. GRIBOV

and

J. NYIRI

CAMBRIDGE UNIVERSITY PRESS
Cambridge, New York, Melbourne, Madrid, Cape Town, Singapore, São Paulo

Cambridge University Press
The Edinburgh Building, Cambridge CB2 2RU, UK

Published in the United States of America by Cambridge University Press, New York

www.cambridge.org
Information on this title: www.cambridge.org/9780521662284

First published 2001
This digitally printed first paperback version 2005

A catalogue record for this publication is available from the British Library

Library of Congress Cataloguing in Publication data

Gribov, V. N. (Vladimir Naumovich)
Quantum electrodynamics : Gribov lectures on theoretical physics / V. N. Gribov, J. Nyiri.
p.   cm. – (Cambridge monographs on particle physics, nuclear physics, and cosmology; 13)
Includes bibliographical references and index.
ISBN 0 521 66228 1
1. Quantum electrodynamics I. Nyiri, J. (Julia), 1939– II. Title. III. Series.
QC680.G75   2000
530.14′33–dc21   99-057211

ISBN-13 978-0-521-66228-4 hardback
ISBN-10 0-521-66228-1 hardback

ISBN-13 978-0-521-67569-7 paperback
ISBN-10 0-521-67569-3 paperback

Contents

Foreword

The idea of this book is to present the theory of quantum electrodynamics in the shortest and clearest way for applied use. At the same time it may serve as a general introduction to relativistic quantum field theory within the approach based on Green functions and the Feynman diagram technique.

The book is largely based on V. N. Gribov's lectures given in Leningrad (St Petersburg) in the early 1970s. The original lecture notes were collected and prepared by V. Fyodorov in 1974.

We were planning several modifications to the work. In particular, Gribov intended to include discussion of his new ideas about the structure of the theory at short distances, the problem he had been working on during his last few years. His death on 13 August 1997 prevented this, and I decided to stay as close as possible to the version completed by early 1997 and already checked by him.

In preparing the book, I got invaluable help from many of our friends and colleagues. I would like to express my gratitude to those who read, commented on, and provided suggestions for improving the manuscript, especially to A. Frenkel. I would also like to thank C. Ewerz and especially Gy. Kluge for their help in preparing the figures.

I am deeply indebted to I. Khriplovich and, most of all, to Gribov's former students, Yu. Dokshitzer, M. Eides and M. Strikman. They performed the enormous work of checking the manuscript by going meticulously through the whole book several times. They compared the text to their own notes taken at Gribov's university courses and restored the Gribov lectures as fully as possible. They found and corrected inconsistencies and errors. It was more than mere scientific editing. Among their objectives was to preserve in the English text the unique style of Gribov the lecturer, a style that is remembered by his disciples and colleagues with admiration.

J. Nyiri
Budapest

1
Particles and their interactions in relativistic quantum mechanics

There are different roads to quantum electrodynamics and to relativistic quantum field theory in general. Three main approaches are those based on

(1) operator secondary quantization technique,

(2) functional integral and

(3) Feynman diagrams.

We shall use the last as physically the most transparent.

1.1 The propagator

In quantum mechanics, the motion of a particle is described by the wave function $\Psi(\mathbf{r}, t)$ which determines the probability amplitudes of all physical processes and satisfies the Schrödinger equation*

$$i\frac{\partial \Psi}{\partial t} = H\Psi. \tag{1.1}$$

The wave function depends on the initial conditions. It is this dependence that makes the notion of wave function inconvenient to use: different wave functions can correspond to essentially the same process. Can one develop a more universal description of physical processes?

Let us introduce the function $K(\mathbf{r}_2, t_2; \mathbf{r}_1, t_1)$, which is called the propagator. Suppose that at time t_1 a particle is placed at a point $\mathbf{r}_1$. We

---

* We use the system of units with $\hbar = c = 1$. Choosing [cm] as the unit of length, these two conditions fix the unit of time [cm] and the unit of mass [cm^{-1}] as well. Indeed, the Compton wavelength of a particle of mass m is $\lambda = \hbar/mc$, i.e. $\lambda = 1/m$ in our case; $t = 1$ cm corresponds to the time which is needed for the light to travel a distance of 1 cm while $m = 1$ cm^{-1} stands for the mass of a (hypothetical) particle, the Compton wavelength of which is $\lambda = 1$ cm.

define $K(\mathbf{r}_2, t_2; \mathbf{r}_1, t_1)$ as the probability amplitude to find this particle at time t_2 at the point $\mathbf{r}_2$. The propagator is a function of four rather than two variables. (This is the price we have to pay for eliminating the arbitrariness of the initial state wave function.)

By definition, $K(\mathbf{r}_2, t_2; \mathbf{r}_1, t_1)$ for $t_2 > t_1$ has to satisfy the Schrödinger equation (1.1), since K is essentially the wave function,

$$K(\mathbf{r}_2, t_2; \mathbf{r}_1, t_1) = \Psi(\mathbf{r}_2, t_2),$$

but with a specific initial condition

$$K(\mathbf{r}_2, t_2; \mathbf{r}_1, t_1) = \Psi(\mathbf{r}_2, t_2)\Big|_{t_2=t_1} = \delta(\mathbf{r}_2 - \mathbf{r}_1). \qquad (1.2)$$

The latter means that at time t_1 the particle was at the point $\mathbf{r}_1$. The knowledge of function K allows us to solve the Cauchy problem for equation (1.1), i.e. to find the wave function of the particle with an arbitrary initial condition $\varphi_{t_1}(\mathbf{r}_1)$:

$$\varphi(\mathbf{r}_2, t_2) = \int K(\mathbf{r}_2, t_2; \mathbf{r}_1, t_1)\, \varphi_{t_1}(\mathbf{r}_1)\, d^3 r_1. \qquad (1.3)$$

The function $\varphi(\mathbf{r}_2, t_2)$ is indeed a solution of (1.1), since the propagator K is a solution of this equation. Moreover, due to (1.2) it also satisfies the initial condition

$$\varphi(\mathbf{r}_2, t_2)\Big|_{t_2=t_1} = \varphi_{t_1}(\mathbf{r}_2).$$

Equation (1.3) means that the probability amplitude to find the particle at the point $\mathbf{r}_2$ at time t_2 is the product of the transition amplitude from $(\mathbf{r}_1, t_1)$ to $(\mathbf{r}_2, t_2)$ and the probability amplitude for the particle to be at time t_1 at the point $\mathbf{r}_1$.

Having a complete orthonormal set of solutions of the stationary Schrödinger equation

$$H\, \Psi_n(\mathbf{r}, t) = E_n \Psi_n(\mathbf{r}, t),$$

we can write the function K as

$$K(\mathbf{r}_2, t_2; \mathbf{r}_1, t_1) = \sum_n \Psi_n(\mathbf{r}_2, t_2)\Psi_n^*(\mathbf{r}_1, t_1). \qquad (1.4)$$

This function, obviously, satisfies equation (1.1) (since $\Psi_n(\mathbf{r}_2, t_2)$ does), while the initial condition (1.2) is satisfied due to completeness of the set of eigenfunctions $\{\Psi_n\}$:

$$K(\mathbf{r}_2, t_1; \mathbf{r}_1, t_1) = \sum_n \Psi_n(\mathbf{r}_2, t_1)\Psi_n^*(\mathbf{r}_1, t_1) = \delta(\mathbf{r}_1 - \mathbf{r}_2).$$

Thus, (1.4) is indeed the propagator.

Let us now determine the propagator for a free particle described by the Hamiltonian

$$H_0 = \frac{\hat{\mathbf{p}}^2}{2m}, \qquad \frac{\hat{\mathbf{p}}^2}{2m}\Psi_n = E_n\Psi_n; \qquad \hat{\mathbf{p}} \equiv -i\frac{d}{d\mathbf{r}}. \qquad (1.5)$$

The solution of (1.5) is

$$\Psi_n(\mathbf{r}, t) = e^{i\mathbf{p}\cdot\mathbf{r} - i\frac{\mathbf{p}^2}{2m}t}, \qquad E_n = \frac{\mathbf{p}^2}{2m}.$$

Since the momentum which determines a state can take arbitrary values, this solution corresponds to the continuous spectrum. Hence, one has to switch from summation to integration over all states in (1.4). As is well known, there are $d^3p/(2\pi)^3$ quantum states in the interval between $\mathbf{p}$ and $\mathbf{p} + d\mathbf{p}$, and therefore one has to substitute $\sum_n \to \int d^3p/(2\pi)^3$ in (1.4). Consequently, for a free particle we obtain

$$\begin{aligned} K_0(\mathbf{r}_2, t_2; \mathbf{r}_1, t_1) &= \int \frac{d^3p}{(2\pi)^3} e^{i\mathbf{p}\cdot\mathbf{r}_2 - i\frac{\mathbf{p}^2}{2m}t_2} e^{-i\mathbf{p}\cdot\mathbf{r}_1 + i\frac{\mathbf{p}^2}{2m}t_1} \\ &= \int \frac{d^3p}{(2\pi)^3} e^{i\mathbf{p}\cdot(\mathbf{r}_2 - \mathbf{r}_1) - i\frac{\mathbf{p}^2}{2m}(t_2 - t_1)}. \end{aligned} \qquad (1.6)$$

It is easy to see that K_0 satisfies both equation (1.1) and the correct initial condition

$$K_0(\mathbf{r}_2, t_1; \mathbf{r}_1, t_1) = \int \frac{d^3p}{(2\pi)^3} e^{i\mathbf{p}\cdot(\mathbf{r}_2 - \mathbf{r}_1)} = \delta(\mathbf{r}_2 - \mathbf{r}_1).$$

From (1.6) it follows also that K_0 is in fact a function of only relative variables, namely: $K_0 = K_0(\mathbf{r}, t)$, where $\mathbf{r} = \mathbf{r}_2 - \mathbf{r}_1$, $t = t_2 - t_1$. This is not surprising, since, if space and time are homogeneous, for a free particle the transition amplitude between $(\mathbf{r}_1, t_1)$ and $(\mathbf{r}_2, t_2)$ must be independent of the absolute position in space and the absolute moment in time.

The integral (1.6) can be calculated explicitly:

$$K_0(\mathbf{r}, t) = \int \frac{d^3p}{(2\pi)^3} e^{i\mathbf{p}\cdot\mathbf{r} - i\frac{\mathbf{p}^2}{2m}t} = \left(\frac{2m}{i\pi t}\right)^{\frac{3}{2}} e^{i\frac{r^2 m}{2t}}.$$

We represent the propagator for a free particle by the line

$$\overline{\phantom{\mathbf{r}_1, t_1 \qquad\qquad\qquad\qquad \mathbf{r}_2, t_2}}$$
$$\mathbf{r}_1, t_1 \qquad\qquad\qquad\qquad\qquad\qquad \mathbf{r}_2, t_2$$

Suppose now that a particle is moving in an external field described by the potential $V(\mathbf{r}, t)$. Let us consider the amplitude which corresponds to the transition of the particle from $(\mathbf{r}_1, t_1)$ to $(\mathbf{r}_2, t_2)$. In this case the following processes are possible:

(1) The particle reaches $(\mathbf{r}_2, t_2)$ without interaction with the external field

$$\overline{\underset{\mathbf{r}_1, t_1}{}\underset{\mathbf{r}_2, t_2}{}} \qquad \begin{array}{c} K_0(\mathbf{r}_2 - \mathbf{r}_1; t_2 - t_1) \\ t_2 > t_1 \end{array} \qquad (1.7)$$

(2) The particle propagates freely up to a point $(\mathbf{r}', t')$ where it interacts with the external field. After this, it continues to propagate freely to $(\mathbf{r}_2, t_2)$. This process can be represented graphically as

$$\overline{\underset{\mathbf{r}_1, t_1}{}\underset{\mathbf{r}'t'}{}\underset{\mathbf{r}_2, t_2}{}} \qquad (1.8)$$

To find the amplitude of this process, let us turn to the Schrödinger equation for a particle in an external field:

$$i\frac{\partial \Psi}{\partial t} = H_0 \Psi + V \Psi.$$

During a small time interval δt the wave function changes by

$$\delta \Psi = -iH_0 \Psi\, \delta t - iV \Psi\, \delta t.$$

The first term on the right-hand side of this equation corresponds to the change of the wave function for free motion which has already been taken into account in (1.7). This means that the interaction with the external field leads to the change

$$\delta_V \Psi = -iV \Psi\, \delta t$$

of the wave function. Based on this observation, we can guess the answer for the amplitude of the process (1.8):

$$K_1(\mathbf{r}_2, t_2; \mathbf{r}_1, t_1) = \int K_0(\mathbf{r}_2 - \mathbf{r}'; t_2 - t')[-iV(\mathbf{r}', t')]K_0(\mathbf{r}' - \mathbf{r}_1; t' - t_1)\, d^3 r'\, dt',$$

$$t_1 < t' < t_2. \qquad (1.9)$$

The integration in (1.9) corresponds to the summation of the amplitudes over all possible positions of the interaction point $(\mathbf{r}', t')$.

(3) The particle interacts twice – at points $(\mathbf{r}', t')$ and $(\mathbf{r}'', t'')$ – with the external field:

$$\overline{\underset{\mathbf{r}_1, t_1}{}\underset{\mathbf{r}', t'}{}\underset{\mathbf{r}'', t''}{}\underset{\mathbf{r}_2, t_2}{}}$$

Similarly to (1.9), we shall write for the amplitude of this process

$$K_2(\mathbf{r}_2, t_2; \mathbf{r}_1, t_1) = \int K_0(\mathbf{r}_2 - \mathbf{r}''; t_2 - t'')\,[-iV(\mathbf{r}'', t'')]\,K_0(\mathbf{r}'' - \mathbf{r}'; t'' - t')$$

$$\times [-iV(\mathbf{r}', t')]\,K_0(\mathbf{r}' - \mathbf{r}_1; t' - t_1)\, d^3 r''\, d^3 r'\, dt''\, dt', \qquad (1.10)$$

$$t_1 < t' < t'' < t_2.$$

It is straightforward to write similar expressions for three or more interactions. We obtain the total transition amplitude $K(\mathbf{r}_2, t_2; \mathbf{r}_1, t_1)$ as a series of amplitudes K_n with n interactions with the external field:

$$K(\mathbf{r}_2, t_2; \mathbf{r}_1, t_1) = \sum_{n=0}^{\infty} K_n(\mathbf{r}_2, t_2; \mathbf{r}_1, t_1). \qquad (1.11)$$

We need to show that the function K so constructed is, indeed, the propagator of a particle in the external field.

1.2 The Green function

Working with the functions K_n, we always have to take care of ordering the successive interaction times. To avoid this inconvenience, we can introduce a new function G

$$
\begin{aligned}
G(\mathbf{r}_2, t_2; \mathbf{r}_1, t_1) &= \theta(t_2 - t_1) \cdot K(\mathbf{r}_2, t_2; \mathbf{r}_1, t_1); && (1.12) \\
G_0(\mathbf{r}_2, t_2; \mathbf{r}_1, t_1) &= \theta(t_2 - t_1) \cdot K_0(\mathbf{r}_2 - \mathbf{r}_1; t_2 - t_1),
\end{aligned}
$$

where

$$\theta(t) = \begin{cases} 1 & t > 0, \\ 0 & t < 0. \end{cases}$$

The function G is called the Green function. The integrals (1.9), (1.10) with G_0 substituted for the free propagators K_0 remain the same, but the step-function θ included in the definition of G ensures the correct time ordering automatically.

Let us now try to find the equation that the Green function satisfies. Acting on G with the operator $i\partial/\partial t - H(\mathbf{r}, t)$, we get

$$
\begin{aligned}
\left[i\frac{\partial}{\partial t_2} - H(\mathbf{r}_2, t_2)\right] G(\mathbf{r}_2, t_2; \mathbf{r}_1, t_1) &= K(\mathbf{r}_2, t_2; \mathbf{r}_1, t_1)\, i\frac{d}{dt}\theta(t_2 - t_1) \\
&= i\delta(\mathbf{r}_2 - \mathbf{r}_1)\delta(t_2 - t_1),
\end{aligned}
$$

if K obeys the Schrödinger equation. In the above derivation we have used that the operator $H(\mathbf{r}, t)$ does not contain time derivatives, the propagator K satisfies (1.2) and the derivative of the step-function $\theta(t)$ gives

$$\frac{d}{dt}\theta(t) = \delta(t).$$

Thus, unlike the propagator, the Green function satisfies the *inhomogeneous* equation:

$$\left[i\frac{\partial}{\partial t_2} - H(\mathbf{r}_2, t_2)\right] G(\mathbf{r}_2, t_2; \mathbf{r}_1, t_1) = i\delta(\mathbf{r}_2 - \mathbf{r}_1)\delta(t_2 - t_1). \qquad (1.13)$$

Let us show now that the total Green function can be obtained as a series

$$G(\mathbf{r}_2, t_2; \mathbf{r}_1, t_1) = \sum_{n=0}^{\infty} G_n(\mathbf{r}_2, t_2; \mathbf{r}_1, t_1), \qquad (1.14)$$

where

$$G_n(\mathbf{r}_2, t_2; \mathbf{r}_1, t_1) \equiv \theta(t_2 - t_1) \cdot K_n(\mathbf{r}_2, t_2; \mathbf{r}_1, t_1).$$

We need to demonstrate that the function G so defined satisfies (1.13).

From now on we will associate each diagram with a respective Green function, for example

$$\Longrightarrow \quad G_0(\mathbf{r}_2, t_2; \mathbf{r}_1, t_1),$$

$$\Longrightarrow \quad G_1(\mathbf{r}_2, t_2; \mathbf{r}_1, t_1)$$

$$= \int G_0(\mathbf{r}_2, t_2; \mathbf{r}', t') \left[-iV(\mathbf{r}', t') \right] G_0(\mathbf{r}', t'; \mathbf{r}_1, t_1) \, dt' d^3 r', \qquad (1.15)$$

etc.

Representing the total Green function by a bold line, we write (1.14) in the graphical form:

All the diagrams on the right-hand side, starting from the second term, have the following structure,

and contain the graph

If we extract this graph as a common factor, the sum of the remaining diagrams again gives the complete Green function

Hence, we can write

$$(1.16)$$

Relation (1.16) is nothing but a graphical equation for the Green function which corresponds to the integral equation

$$G(\mathbf{r}_2, t_2; \mathbf{r}_1, t_1) = G_0(\mathbf{r}_2, t_2; \mathbf{r}_1, t_1) \tag{1.17}$$
$$+ \int G_0(\mathbf{r}_2, t_2; \mathbf{r}', t') \left[-iV(\mathbf{r}', t') \right] G(\mathbf{r}', t'; \mathbf{r}_1, t_1) \, d^3r' dt' \, .$$

To prove that it is equivalent to the differential equation (1.13) we apply the free Schrödinger operator to (1.17):

$$\left[i\frac{\partial}{\partial t_2} - H_0(\mathbf{r}_2, t_2) \right] G(\mathbf{r}_2, t_2; \mathbf{r}_1, t_1) = i\delta(\mathbf{r}_2 - \mathbf{r}_1)\delta(t_2 - t_1)$$
$$+ \int i\delta(\mathbf{r}_2 - \mathbf{r}')\delta(t_2 - t') \left[-iV(\mathbf{r}', t') \right] G(\mathbf{r}', t'; \mathbf{r}_1, t_1) \, d^3r' dt'$$
$$= i\delta(\mathbf{r}_2 - \mathbf{r}_1)\delta(t_2 - t_1) + V(\mathbf{r}_2, t_2) \, G(\mathbf{r}_2, t_2; \mathbf{r}_1, t_1) \, .$$

Moving the second term from the right-hand side to the left-hand side of this equation, we get exactly (1.13). The Green function G is defined unambiguously as a solution of the inhomogeneous differential equation (1.13) (or the inhomogeneous integral equation (1.17)) with the initial condition $G(\mathbf{r}_2, t_2; \mathbf{r}_1, t_1) = 0$ for $t_2 < t_1$. Note that in the case of the integral equation, this condition is automatically satisfied by the iterative (perturbative) solution: the exact Green function G vanishes for $t_2 < t_1$ because the free Green function G_0 does.

We conclude that the function G constructed according to the prescription (1.14) is indeed the Green function of the Schrödinger equation for a particle in an external field. Using (1.12) this allows us to complete the proof that the function K in (1.11) is the corresponding propagator.

The graphs introduced above are, in fact, Feynman diagrams for the scattering of a particle in an external field in the non-relativistic case.

It is worthwhile to notice that the space and time variables enter on equal footing in equation (1.13) for the Green function, as well as in the integrals for G_n. This symmetry will make the Green function our main tool when we turn to constructing the relativistic theory.

1.2.1 The Green function for a system of particles

The Green function for two or more particles can be constructed in the same way. Consider, for example, two free particles. Their motion can be described as

$$\overline{\quad \mathbf{r}_1, t_1 \qquad\qquad \mathbf{r}_1', t_1' \quad}$$

$$\overline{\quad \mathbf{r}_2, t_2 \qquad\qquad \mathbf{r}_2', t_2' \quad}$$

Since these particles are moving independently of each other, the Green function in this case is simply the product of the one-particle Green functions:

$$G_0(\mathbf{r}'_2, \mathbf{r}'_1, t'_2, t'_1; \mathbf{r}_2, \mathbf{r}_1, t_2, t_1) = G_0(\mathbf{r}'_1 - \mathbf{r}_1, t'_1 - t_1)G_0(\mathbf{r}'_2 - \mathbf{r}_2, t'_2 - t_2).$$

The simplest diagram which takes into account the interaction between two particles is

$$\text{(1.18)}$$

The dashed line corresponds to a single interaction between the particles. Similarly to the case of one particle in an external field, we ascribe to this diagram the factor $[-iV(\mathbf{x}_2 - \mathbf{x}_1, \tau_2 - \tau_1)]$, with V the interaction potential. For G_1 we obtain

$$G_1 = \int G_0(\mathbf{r}'_1, t'_1; \mathbf{x}_1, \tau_1)[-iV(\mathbf{x}_2 - \mathbf{x}_1; \tau_2 - \tau_1)]G_0(\mathbf{x}_1, \tau_1; \mathbf{r}_1, t_1)$$
$$\times G_0(\mathbf{r}'_2, t'_2; \mathbf{x}_2, \tau_2)G_0(\mathbf{x}_2, \tau_2; \mathbf{r}_2, t_2)\, d^3x_1\, d^3x_2\, d\tau_1\, d\tau_2. \qquad \text{(1.19)}$$

Unlike the case of one particle in an external field, the potential in (1.18) describes an interaction between two particles, and it enters the respective analytic expression in (1.19) only once. A justification of the prescription (1.19) will be presented in Section 1.7. (Note that in non-relativistic theory the interaction is instantaneous, so that actually $V(\mathbf{x}_2 - \mathbf{x}_1; \tau_2 - \tau_1) = \delta(\tau_2 - \tau_1)\, V(\mathbf{x}_2 - \mathbf{x}_1)$.)

1.2.2 The momentum representation

We now return to the case of a particle in an external field. Usually it is very instructive to work in momentum space. We shall carry out a transformation to the momentum representation in a way which allows us to preserve the formal symmetry between space and time variables. This symmetry will be useful later, when generalizing the theory to the relativistic case.

The Green function of the free particle is

$$G_0(\mathbf{r}, t) = \int \frac{d^3p}{(2\pi)^3}\, e^{i\mathbf{p}\cdot\mathbf{r} - i\frac{\mathbf{p}^2}{2m}t}\, \theta(t) \qquad \text{(1.20)}$$

(see (1.6) and (1.12)). The variables t and $\mathbf{r}$ enter this expression in a non-symmetric way. The symmetry can be restored, however, if we write

$$G_0(\mathbf{r}, t) = \int \frac{d^4p}{(2\pi)^4 i} \, G_0(\mathbf{p}, p_0) \, e^{i\mathbf{p}\cdot\mathbf{r} - i p_0 t}, \tag{1.21}$$

where $\mathbf{r}$ and t enter on equal footing as do $\mathbf{p}$, p_0. Here the Green function in the momentum representation is

$$G_0(\mathbf{p}, p_0) = \frac{1}{\frac{\mathbf{p}^2}{2m} - p_0 - i\varepsilon}, \tag{1.22}$$

where ε is an arbitrarily small positive number. Thus,

$$G_0(\mathbf{r}, t) = \int \frac{d^3p \, dp_0}{(2\pi)^4 i} \, \frac{1}{\frac{\mathbf{p}^2}{2m} - p_0 - i\varepsilon} \, e^{i\mathbf{p}\cdot\mathbf{r} - i p_0 t}. \tag{1.23}$$

Let us show now that (1.23) and (1.20) are equivalent. For this purpose, integrate (1.23) over p_0. The integrand has a simple pole in the lower half-plane at

$$p_0 = \frac{\mathbf{p}^2}{2m} - i\varepsilon.$$

If we had $\varepsilon = 0$, the pole would be located on the real axis and the integral would not make sense.

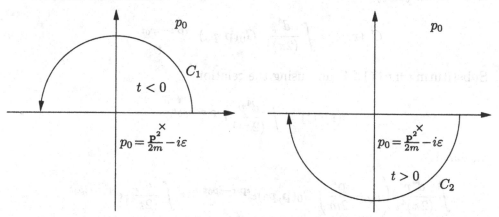

If $t < 0$, the contour of integration can be closed in the upper half-plane. Since in this case there are no poles inside the contour C_1, the contour integral vanishes. At the same time the integral over the upper half-circle is zero due to the Jordan lemma[†] and this leads to zero for the integral (1.23).

[†] The Jordan lemma is proved in any textbook on mathematical physics, see e.g. G. B. Arfken and J. Weber, *Mathematical Methods for Physicists*, Academic Press, 1995

Consider $t > 0$ and close the contour of integration in the lower half-plane. This time the integral over the *lower* half-circle is zero, giving

$$(1.23) = \int_{C_2} \frac{d^3p\, dp_0}{(2\pi)^4 i} \frac{1}{\frac{\mathbf{p}^2}{2m} - p_0 - i\varepsilon} e^{i\mathbf{p}\cdot\mathbf{r} - ip_0 t} = -2\pi i \operatorname{Res}\Big|_{p_0} .$$

Here $\operatorname{Res}\big|_{p_0}$ is the residue of the integrand at the point $p_0 = \mathbf{p}^2/2m - i\varepsilon$. Taking the $\varepsilon \to 0$ limit we get

$$(1.23) = 2\pi i \int \frac{d^3p}{(2\pi)^4 i} e^{i\mathbf{p}\cdot\mathbf{r} - i\frac{\mathbf{p}^2}{2m} t},$$

which is exactly the expression (1.20) for $t > 0$. Hence, we have proved that (1.20) and (1.23) coincide.

In (1.23) we guessed the expression for the function $G_0(\mathbf{p}, p_0)$. Let us obtain (1.22) straight from the Schrödinger equation. So, we are looking for the solution of the equation

$$\left(i\frac{\partial}{\partial t} + \frac{\mathbf{\nabla}^2}{2m} \right) G_0(\mathbf{r}, t) = i\delta(\mathbf{r})\delta(t) \qquad (1.24)$$

in the form (1.21):

$$G_0(\mathbf{r}, t) = \int \frac{d^4p}{(2\pi)^4 i} \, G_0(\mathbf{p}, p_0)\, e^{i\mathbf{p}\cdot\mathbf{r} - ip_0 t} .$$

Substituting into (1.24) and using the relation

$$\delta(\mathbf{r})\delta(t) = \int \frac{d^4p}{(2\pi)^4} \, e^{i\mathbf{p}\cdot\mathbf{r} - ip_0 t} ,$$

we have

$$\int \frac{d^4p}{(2\pi)^4 i} \left(p_0 - \frac{\mathbf{p}^2}{2m} \right) G_0(\mathbf{p}, p_0) e^{i\mathbf{p}\cdot\mathbf{r} - ip_0 t} = i \int \frac{d^4p}{(2\pi)^4} e^{i\mathbf{p}\cdot\mathbf{r} - ip_0 t} ,$$

which, after introducing the infinitesimal quantity $-i\varepsilon$ to satisfy the condition $G_0(\mathbf{r}, t) = 0$ for $t < 0$, results in (1.22).

Let us introduce the momentum representation for the external potential:

$$V(\mathbf{r}, t) = \int \frac{d^4q}{(2\pi)^4} e^{i\mathbf{q}\cdot\mathbf{r} - iq_0 t} V(q) ,$$

$$V(q) = \int d^3r\, dt\, e^{-i\mathbf{q}\cdot\mathbf{r} + iq_0 t} V(\mathbf{r}, t) ,$$

$$(1.25)$$

where $q = (q_0, \mathbf{q})$. By substituting (1.21) and (1.25) into the expression (1.15) for $G_1(\mathbf{r}_2, t_2; \mathbf{r}_1, t_1)$ we can describe the process

$$
\frac{\quad G_0 \qquad\quad -iV \qquad\quad G_0 \quad}{\mathbf{r}_1, t_1 \qquad\qquad \mathbf{r}', t' \qquad\qquad \mathbf{r}_2, t_2}
$$

in the form

$$
G_1(\mathbf{r}_2, t_2; \mathbf{r}_1, t_1) = \int d^3r' dt' \underbrace{\int \frac{d^4 p_2}{(2\pi)^4 i} G_0(p_2) e^{i\mathbf{p}_2 \cdot (\mathbf{r}_2 - \mathbf{r}') - ip_{20}(t_2 - t')}}_{G_0(\mathbf{r}_2, t_2; \mathbf{r}', t')}
$$

$$
\times \underbrace{(-i) \int \frac{d^4 q}{(2\pi)^4} V(q) e^{i\mathbf{q} \cdot \mathbf{r}' - iq_0 t'}}_{-iV(\mathbf{r}', t')} \underbrace{\int \frac{d^4 p_1}{(2\pi)^4 i} G_0(p_1) e^{i\mathbf{p}_1 \cdot (\mathbf{r}' - \mathbf{r}_1) - ip_{10}(t' - t_1)}}_{G_0(\mathbf{r}', t'; \mathbf{r}_1, t_1)}
$$

$$
= \int \frac{d^4 p_2\, d^4 q\, d^4 p_1}{(2\pi)^{12} i} \underbrace{\int d^3 r' dt' e^{i(-\mathbf{p}_2 + \mathbf{q} + \mathbf{p}_1) \cdot \mathbf{r}'} e^{i(p_{20} - q_0 - p_{10}) t'}}_{(2\pi)^4 \delta^4 (p_1 + q - p_2)}
$$

$$
\times\ e^{i\mathbf{p}_2 \cdot \mathbf{r}_2 - ip_{20} t_2} e^{-i\mathbf{p}_1 \cdot \mathbf{r}_1 + ip_{10} t_1} G_0(p_2) \left[-V(q) \right] G_0(p_1).
$$

The integration over $\mathbf{r}'$ and t' leads to the δ-function which ensures the energy–momentum conservation. Integrating over q, we finally obtain

$$
G_1(\mathbf{r}_2, t_2; \mathbf{r}_1, t_1) = \int \frac{d^4 p_1 d^4 p_2}{(2\pi)^8 i} e^{-i\mathbf{p}_1 \cdot \mathbf{r}_1 + ip_{10} t_1} e^{i\mathbf{p}_2 \cdot \mathbf{r}_2 - ip_{20} t_2}
$$
$$
\times\ G_0(p_2) \left[-V(p_2 - p_1) \right] G_0(p_1). \tag{1.26}
$$

Hence, as a result of the interaction, the first correction to the Green function of the free particle will no longer be a function of only the differences $\mathbf{r} = \mathbf{r}_2 - \mathbf{r}_1$ and $t = t_2 - t_1$. G_0 can be rewritten in a similar form:

$$
G_0(\mathbf{r}, t) = \int \frac{d^4 p_1 d^4 p_2}{(2\pi)^4 i} e^{-i\mathbf{p}_1 \cdot \mathbf{r}_1 + ip_{10} t_1} e^{i\mathbf{p}_2 \cdot \mathbf{r}_2 - ip_{20} t_2} \delta(p_1 - p_2) G_0(p_1). \tag{1.27}
$$

Let us now introduce the exact Green function $G(p_1, p_2)$ in momentum representation:

$$
G(\mathbf{r}_2, t_2; \mathbf{r}_1, t_1) = \int \frac{d^4 p_1 d^4 p_2}{(2\pi)^8 i} e^{i(\mathbf{p}_2 \cdot \mathbf{r}_2 - p_{20} t_2)} e^{-i(\mathbf{p}_1 \cdot \mathbf{r}_1 - p_{10} t_1)} G(p_1, p_2).
$$
$$
\tag{1.28}
$$

Taking into account (1.27) and (1.26), we obtain

$$
G(p_1, p_2) = (2\pi)^4 \delta(p_1 - p_2) G_0(p_1) + G_0(p_2)[-V(p_2 - p_1)] G_0(p_1) + \cdots. \tag{1.29}
$$

Graphically this will look like

$$\frac{G(p)}{p_1 \quad p_2} = \frac{(2\pi)^4 \delta(p_1 - p_2) G_0(p_2)}{p_1 \quad p_2} + \frac{G_0(p_1) - V \quad G_0(p_2)}{p_1 \quad p_2} + \cdots$$

The expression for $G_1(p_1, p_2)$ given by the second term in (1.29) corresponds to the first Born approximation in non-relativistic quantum mechanics ($-V(p_2 - p_1)$ is the scattering amplitude in this approximation).

One can repeat this procedure for the diagram

$$\begin{array}{ccccc} p_1 & & p' & & p_2 \\ \mathbf{r_1}, t_1 & \mathbf{r}', \mathbf{t}' & & \mathbf{r}'', \mathbf{t}'' & \mathbf{r_2}, t_2 \end{array}$$

and get for the next term in the perturbative series (1.29) for the Green function

$$G_2(p_1, p_2) = G_0(p_1) \cdot \int \frac{d^4 p'}{(2\pi)^4} V(p' - p_1) G_0(p') V(p_2 - p') \cdot G_0(p_2). \quad (1.30)$$

This corresponds to the second Born approximation, and the integration over the momenta here is equivalent to the summation over the intermediate states in the standard quantum mechanical approach.

The general rules for constructing the Green functions G_n that correspond to diagrams

$$\begin{array}{ccccc} p_1 & & p' & p'' & p_2 \\ & -V(p' - p_1) & -V(p'' - p') & -V(p_2 - p'') & \end{array} + \cdots$$

can be formulated in a similar way. Namely, every line corresponds to a free Green function $G_0(p)$, every vertex corresponds to $(-V)$, and integrations with the weight $d^4 p/(2\pi)^4$ have to be carried out over all momenta of the intermediate lines.

1.2.3 Virtual particles

Our perturbation theory differs from the usual one in the following respect. The non-relativistic quantum mechanical expressions look as if the energy was not conserved. Consider, for example, the stationary scattering problem, $V(\mathbf{r}, t) = V(\mathbf{r})$. In the second order in V one writes for the scattering amplitude

$$f^{(2)}_{\mathbf{p_1} \to \mathbf{p_2}} = \sum_{\mathbf{k}} \frac{V_{\mathbf{p_2 k}} V_{\mathbf{k p_1}}}{\frac{k^2}{2m} - \frac{p_1^2}{2m}}, \qquad V_{\mathbf{kp}} = \int d^3 r \, e^{i(\mathbf{p} - \mathbf{k}) \cdot \mathbf{r}} V(\mathbf{r}), \qquad (1.31)$$

where $V_{\mathbf{kp}}$ is the matrix element of the potential between the free particle states with three-momenta $\mathbf{p}$ and $\mathbf{k}$. Although the energies of the initial-

and final-state particles are the same, $\mathbf{p}_1^2/2m = \mathbf{p}_2^2/2m = E$, the intermediate state energies $E_\mathbf{k} = \mathbf{k}^2/2m$ are arbitrary and, generally speaking, different from E.

The expression (1.31) is actually contained in our G_2. Indeed, using that in the stationary case $V(q) = 2\pi\delta(q_0)V(\mathbf{q})$, we can represent the integral in (1.30) in the form

$$\sum_{\mathbf{p}'} \int \frac{dp_0'}{2\pi} V(p_2-p') G_0(p')V(p'-p_1) \Longrightarrow \sum_{\mathbf{p}'} \frac{V(\mathbf{p}_2-\mathbf{p}')V(\mathbf{p}'-\mathbf{p}_1)}{\frac{\mathbf{p}'^2}{2m} - p_0'}\Bigg|_{p_{10}=p_0'=p_{20}}$$

This becomes identical to the non-relativistic amplitude (1.31) if we take the real external particle with $p_{10} = \mathbf{p}_1^2/2m$, and substitute $\mathbf{p}_1^2/2m$ for $p_0' = p_0$.

What remains different is the interpretation. Within the framework of our new perturbation theory the *energy* defined as the zero-component of the four-momentum $p = (p_0, \mathbf{p})$ remains conserved at all stages of the process, $p_{10} = p_0' = p_{20}$. The intermediate particle p', however, is not *real* because its energy and three-momentum do not satisfy the relation characterizing a free physical state, $p_0' \neq \mathbf{p}'^2/2m$. It is a *virtual* particle (or particle in a virtual state).

1.3 The scattering amplitude

1.3.1 How to calculate physical observables

Let us calculate, for example, the scattering amplitude. The initial state in the remote past is described by the wave function $\Psi_i(\mathbf{r}, t_1)$. As a result of the interaction, the particle at finite times $t > t_1$ is described by the wave function $\Psi(\mathbf{r}, t)$. This wave function contains information about the interaction and 'remembers' the initial state. What we access experimentally is the probability amplitude to find the particle in the remote future in a given state $\Psi_f(\mathbf{r}, t)$:

$$\int \Psi_f^*(\mathbf{r}, t)\Psi(\mathbf{r}, t)\, d^3r \,.$$

This expression can be simplified with the help of the propagator. Since

$$\Psi(\mathbf{r}, t) = \int K(\mathbf{r}, t; \mathbf{r}', t')\Psi_i(\mathbf{r}', t')\, d^3r' \,,$$

the transition amplitude i→f (or the matrix element of the scattering matrix S) has the form

$$S_{fi} = \int \Psi_f^*(\mathbf{r}, t)\Psi(\mathbf{r}, t)d^3r = \int \Psi_f^*(\mathbf{r}, t)K(\mathbf{r}, t; \mathbf{r}', t')\Psi_i(\mathbf{r}', t')\, d^3r d^3r' \,,$$

14   *1 Particles and their interactions*

where $t \to \infty$, $t' \to -\infty$. Finally, substituting the function G instead of K, we get

$$S_{\mathrm{fi}} = \int \Psi_{\mathrm{f}}^*(\mathbf{r},t)\, G(\mathbf{r},t;\mathbf{r}',t')\, \Psi_{\mathrm{i}}(\mathbf{r}',t')\, d^3r d^3r'. \tag{1.32}$$

Now calculate (1.32) for a real process. Suppose that a particle with momentum $\mathbf{p}_1$ interacts with an external field and, as a result, makes a transition into a state with momentum $\mathbf{p}_2$, i.e.

$$\Psi_{\mathrm{i}} = \Psi_1 = e^{i\mathbf{p}_1\cdot\mathbf{r}-i\frac{\mathbf{p}_1^2}{2m}t}, \qquad \Psi_{\mathrm{f}} = \Psi_2 = e^{i\mathbf{p}_2\cdot\mathbf{r}-i\frac{\mathbf{p}_2^2}{2m}t}.$$

Then

$$S_{\mathbf{p}_2,\mathbf{p}_1} = \int d^3r' d^3r\, e^{-i\mathbf{p}_2\cdot\mathbf{r}+i\frac{\mathbf{p}_2^2}{2m}t}\, e^{i\mathbf{p}_1\cdot\mathbf{r}'-i\frac{\mathbf{p}_1^2}{2m}t'}$$
$$\times \int \frac{d^4p_1' d^4p_2'}{(2\pi)^8 i}\, G(p_1',p_2')\, e^{i\mathbf{p}_2'\cdot\mathbf{r}-ip_{20}'t}\, e^{-i\mathbf{p}_1'\cdot\mathbf{r}'+ip_{10}'t'}, \tag{1.33}$$

where we have used the momentum space representation (1.28) for the Green function $G(\mathbf{r},t;\mathbf{r}',t')$. The integration over r and r' in (1.33) generates the product of delta-functions,

$$(2\pi)^3\delta(\mathbf{p}_2-\mathbf{p}_2')\,(2\pi)^3\delta(\mathbf{p}_1-\mathbf{p}_1'),$$

and we obtain

$$S_{\mathbf{p}_2,\mathbf{p}_1} = \frac{1}{(2\pi)^2 i}\int dp_{10} dp_{20}\, e^{it(\frac{\mathbf{p}_2^2}{2m}-p_{20})}\, e^{-it'(\frac{\mathbf{p}_1^2}{2m}-p_{10})}\, G(p_1,p_2). \tag{1.34}$$

‡ Recall now the expansion of $G(p_1,p_2)$ into the series (1.29):

$$G(p_1,p_2) = (2\pi)^4\delta(p_1-p_2)G_0(p_1) + G_0(p_2)\left[-V(p_2-p_1)\right]G_0(p_1)$$
$$+ G_0(p_1)\int \frac{d^4p'}{(2\pi)^4}V(p'-p_1)\,G_0(p')\,V(p_2-p')\,G_0(p_2) + \cdots.$$

Inserting this expression into (1.34), we obtain the first term in the form

$$S_{\mathbf{p}_2,\mathbf{p}_1}^0 = -i(2\pi)^2\delta(\mathbf{p}_1-\mathbf{p}_2)\int dp_{10}\,G_0(\mathbf{p}_1,p_{10})\,e^{i(t-t')\left(\frac{\mathbf{p}_1^2}{2m}-p_{10}\right)}$$
$$= -i(2\pi)^2\delta(\mathbf{p}_1-\mathbf{p}_2)\int \frac{dp_0}{\frac{\mathbf{p}_1^2}{2m}-p_0-i\varepsilon}\,e^{i(t-t')\left(\frac{\mathbf{p}_1^2}{2m}-p_0\right)} \tag{1.35}$$
$$= (2\pi)^3\delta(\mathbf{p}_1-\mathbf{p}_2),$$

‡ Note that in passing to (1.34) we have renamed the integration variables $p_{i0}' \to dp_{i0}$, along with the substitution $\mathbf{p}_i' = \mathbf{p}_i$ due to the delta-functions above. This allows us to keep using the compact four-momentum notation $p_i = (p_{0i},\mathbf{p}_i)$ but does not imply that p_{0i} coincides with the energy of the real external particle, $\mathbf{p}_i^2/2m$.

i.e. there is no scattering in this approximation. In order to derive (1.35), the explicit form (1.22) of the Green function $G_0(\mathbf{p}, p_0)$ has been used. Since $t - t' > 0$, the contour of integration was closed in the lower half-plane, with the integration yielding $2\pi i$.

All the other terms in the expansion of $G(p_1, p_2)$ contain free Green functions sidewise. Therefore, we can write

$$G(p_1, p_2) = (2\pi)^4 \delta(p_1 - p_2) G_0(p_1) + G_0(p_1)\, T(p_1, p_2)\, G_0(p_2), \quad (1.36)$$

where $T(p_1, p_2)$ contains all the internal lines and the integrations over the intermediate momenta.

Let us now calculate the contribution of the second term in (1.36) to the integral (1.34). When $t \to \infty$ and $t' \to -\infty$, the exponential factors in the integrand oscillate rapidly. If the integrand were a smooth function, the integral would turn to zero. However, this is not the case, since there are poles of the free Green functions:

$$G_0(p_1) G_0(p_2) = \frac{1}{\left(\frac{\mathbf{p}_1^2}{2m} - p_{10} - i\varepsilon\right)\left(\frac{\mathbf{p}_2^2}{2m} - p_{20} - i\varepsilon\right)}.$$

Moreover, these are the only poles: the factor $T(p_1, p_2)$ as a function of the external energy variables p_{10}, p_{20} is smoother, because its internal singularities are being integrated over,

$$\int \frac{d^4 p'}{\frac{\mathbf{p}'^2}{2m} - p_0' - i\varepsilon}\, V(p_0' - p_{10}; \mathbf{p}' - \mathbf{p}_1) \cdots.$$

Hence, the integrals in (1.34) can be calculated by residues, and we finally obtain for the transition amplitude

$$S_{\mathbf{p}_2, \mathbf{p}_1} = (2\pi)^3 \delta(\mathbf{p}_2 - \mathbf{p}_1) + i T(p_1, p_2). \quad (1.37)$$

This means that $T(p_1, p_2)$ is the scattering amplitude, and it can be calculated in the following way:

(1) Draw the relevant diagrams:

(2) Write the corresponding Green function according to the rules above:

$$G(p_1, p_2) = (2\pi)^4 \delta(p_1 - p_2) G_0(p_1) + \frac{T(p_1, p_2)}{\left(\frac{\mathbf{p}_2^2}{2m} - p_{20} - i\varepsilon\right)\left(\frac{\mathbf{p}_1^2}{2m} - p_{10} - i\varepsilon\right)}.$$

(3) Throw away the pole factors $G_0(p_1)$ and $G_0(p_2)$.

(4) Take the external line momenta to describe the real particles, that is $p_{10} = \mathbf{p}_1^2 / 2m$, $p_{20} = \mathbf{p}_2^2 / 2m$.

1.3.2 Poles in the scattering amplitude and the bound states

Let us now show that the poles in the scattering amplitude determine the bound state energies. The usual quantum mechanical scattering amplitude is

$$f = -\frac{2m}{4\pi} \int e^{-i\mathbf{p}'\cdot\mathbf{r}} V(\mathbf{r}) \Psi_E(\mathbf{r}) \, d^3r, \qquad \frac{\mathbf{p}'^2}{2m} = E, \qquad (1.38)$$

where $\mathbf{p}'$ is the final particle momentum and $\Psi_E(\mathbf{r})$ is the exact solution of the stationary Schrödinger equation with a given energy E.

For the Green function we have the expression

$$G(\mathbf{r}_2, t_2; \mathbf{r}_1, t_1) = \theta(\tau) \sum_n \Psi_n(\mathbf{r}_2, t_2) \Psi_n^*(\mathbf{r}_1, t_1); \qquad \tau \equiv t_2 - t_1.$$

Consider the Green function with a definite energy:

$$\begin{aligned}
G_E(\mathbf{r}_2, \mathbf{r}_1) &= \int_{-\infty}^{\infty} G(\mathbf{r}_2, \mathbf{r}_1, \tau) \, e^{iE\tau} \, d\tau \\
&= \sum_n \Psi_n(\mathbf{r}_2) \Psi_n^*(\mathbf{r}_1) \int_0^{\infty} d\tau \, e^{i(E-E_n)\tau} \qquad (1.39) \\
&= \frac{1}{i} \sum_n \frac{\Psi_n(\mathbf{r}_2) \Psi_n^*(\mathbf{r}_1)}{E_n - E}.
\end{aligned}$$

This function satisfies the equation

$$(H - E) \, G_E(\mathbf{r}_2, \mathbf{r}_1) = \frac{1}{i} \delta(\mathbf{r}_2 - \mathbf{r}_1).$$

The sign $\sum_n$ in (1.39) implies integration over the continuous spectrum and summation over the states belonging to the discrete spectrum, i.e. the bound states (if any). We see therefore that $G_E(\mathbf{r}_2, \mathbf{r}_1)$ as a function of E has poles at the bound state energies. Let us demonstrate that these very poles show up in the scattering amplitude as well.

With the help of $G_E(\mathbf{r}_1, \mathbf{r}_2)$ one can construct the exact solutions of the stationary Schrödinger equation. In particular, for the incoming particle with momentum $\mathbf{p}$ we write

$$\Psi_E(\mathbf{r}) = e^{i\mathbf{p}\cdot\mathbf{r}} + (-i) \int G_E(\mathbf{r}, \mathbf{r}') \, e^{i\mathbf{p}\cdot\mathbf{r}'} \, V(\mathbf{r}') \, d^3r'.$$

This function indeed satisfies the Schrödinger equation:

$$\begin{aligned}
(H - E)\Psi_E &= (H - E)e^{i\mathbf{p}\cdot\mathbf{r}} - i \int (H - E)G_E(\mathbf{r}, \mathbf{r}') \, e^{i\mathbf{p}\cdot\mathbf{r}'} V(\mathbf{r}') \, d^3r' \\
&= V(\mathbf{r})e^{i\mathbf{p}\cdot\mathbf{r}} - V(\mathbf{r})e^{i\mathbf{p}\cdot\mathbf{r}} = 0; \qquad \text{for } \frac{\mathbf{p}^2}{2m} = E.
\end{aligned}$$

Then

$$f = - \frac{2m}{4\pi} \int e^{i\mathbf{q}\cdot\mathbf{r}} V(\mathbf{r}) \, d^3r$$

$$+ \frac{2m}{4\pi} i \int e^{-i\mathbf{p}'\cdot\mathbf{r}} V(\mathbf{r}) \, G_E(\mathbf{r},\mathbf{r}') \, e^{i\mathbf{p}\cdot\mathbf{r}'} V(\mathbf{r}') \, d^3r \, d^3r' \qquad (1.40)$$

$$= f_B + \frac{2m}{4\pi} \sum_n \frac{f_{np'} f_{np}^*}{E_n - E},$$

where $\mathbf{q} = \mathbf{p} - \mathbf{p}'$, f_B is the scattering amplitude in the Born approximation, and

$$f_{np} = \int e^{-i\mathbf{p}\cdot\mathbf{r}} V(\mathbf{r}) \, \Psi_n(\mathbf{r}) \, d^3r .$$

We see from (1.40) that the bound states really correspond to the poles of the scattering amplitude.

1.4 The electromagnetic field

Aiming at relativistic quantum field theory it is natural to start by considering the intrinsically relativistic object – the electromagnetic field. So, the first question we set for ourselves is how to construct the quantum mechanics of the photon?

In classical physics one introduces a four-tensor $F_{\mu\nu}(x)$ of the electromagnetic field, the components of which are the strengths of the electric and magnetic fields $\mathbf{E}$ and $\mathbf{H}$ (hereafter, x is a four-vector $x \equiv (\mathbf{x}, t)$).

We write the relativistic invariants in the form

$$x_\mu y_\mu = x_0 y_0 - x_1 y_1 - x_2 y_2 - x_3 y_3 ,$$

making no distinction between the upper and lower indices. We introduce also the *metric tensor* $g_{\mu\nu}$:

$$g_{00} = 1, \quad g_{11} = g_{22} = g_{33} = -1; \qquad g_{\mu\nu} = 0 \quad \text{if} \quad \mu \neq \nu .$$

We use the system of units where the fine structure constant $\alpha \simeq 1/137$ is connected to the unit of charge by the relation

$$e^2 = 4\pi\alpha \simeq \frac{4\pi}{137} .$$

In this system of units the Maxwell equations for $F_{\mu\nu}(x)$ have the form

$$\frac{\partial F_{\nu\mu}}{\partial x_\nu} = j_\mu(x), \qquad (1.41)$$

$$\frac{\partial F_{\mu\nu}}{\partial x_\lambda} + \frac{\partial F_{\nu\lambda}}{\partial x_\mu} + \frac{\partial F_{\lambda\mu}}{\partial x_\nu} = 0 . \qquad (1.42)$$

These are relativistically invariant classical equations of the electromagnetic field. Usually one introduces the potentials $A_\mu(x)$

$$F_{\mu\nu} = \frac{\partial A_\nu}{\partial x_\mu} - \frac{\partial A_\mu}{\partial x_\nu} \qquad (1.43)$$

and (1.42) is valid automatically. There is, however, an ambiguity in the choice of the potentials A_μ since the relation (1.43) does not fix them uniquely. Using this ambiguity one can impose an additional Lorentz condition

$$\frac{\partial A_\mu}{\partial x_\mu} = 0\,.$$

Then (1.41) turns into the wave equation for the potentials:

$$\Box A_\mu(x) \equiv \frac{\partial^2 A_\mu}{\partial x_\nu \partial x_\nu} = j_\mu(x)\,. \qquad (1.44)$$

For the time being we suppress the vector index and consider the d'Alembert equation in empty space,

$$\Box f(x) = 0\,, \qquad (1.45)$$

which describes the propagation of free electromagnetic waves. Let us try to describe the free electromagnetic field quantum mechanically. Although such a description will give nothing new in the free case, it will be necessary for generalization to the case of interaction.

Suppose that the electromagnetic field consists of photons – quantum particles which are described by a certain wave function. The laws of motion should be identical for the free classical field and the corresponding free quantum particle, since the quantum effects begin to manifest themselves only when the influence of a measuring device is not small, i.e. when an interaction is present. Therefore, the free photon wave function we are looking for should satisfy the classical wave equation (1.45).

Note that the classical electromagnetic field is observable since the change in the field after the interaction with the measuring device is negligible. In quantum mechanics the situation is different. Here a particle is described by the wave function $\Psi(x)$ which cannot be measured directly. However, its absolute value squared $|\Psi(x)|^2$ (the probability density in non-relativistic quantum mechanics) determines physical observables and is, in this sense, measurable. The integral of the probability density over the whole three-dimensional space is equal to the probability to find a quantum mechanical particle anywhere in space and turns out to be time-independent,

$$\int |\Psi|^2\, d^3r = \text{const}\,. \qquad (1.46)$$

Conservation of probability is one of the most fundamental principles of quantum mechanics.

We have to construct for the photon a wave function Ψ which admits probabilistic interpretation, i.e. there should exist a probability density ($|\Psi|^2$, or its analogue) with conserved spatial integral, as in (1.46). On the other hand, this Ψ has to satisfy (1.45) which describes propagation of photons with constant velocity c in vacuum. Can we construct the solution of the d'Alembert equation with the necessary property?

Recall how the conservation of the integral (1.46) is derived in non-relativistic quantum mechanics. The wave function Ψ is a complex function which satisfies the Schrödinger equation

$$i\frac{\partial\Psi}{\partial t} = H\Psi, \quad H = -\frac{\nabla^2}{2m}. \tag{1.47}$$

The complex conjugate of (1.47) is

$$-i\frac{\partial\Psi^*}{\partial t} = H^*\Psi^*. \tag{1.48}$$

Multiplying (1.47) by Ψ^*, (1.48) by Ψ and adding the two expressions, we obtain

$$i\frac{\partial}{\partial t}\left(\Psi^*\Psi\right) = \frac{1}{2m}(\Psi\nabla^2\Psi^* - \Psi^*\nabla^2\Psi)$$

or

$$\frac{\partial}{\partial t}\left(\Psi^*\Psi\right) = -\frac{i}{2m}\,\mathrm{div}(\Psi\nabla\Psi^* - \Psi^*\nabla\Psi). \tag{1.49}$$

The equation of continuity (1.49) allows us to interpret $\Psi^*\Psi$ as the probability density, since

$$\Psi^*\Psi > 0 \quad \text{and} \quad \int \Psi^*\Psi \, d^3r = \text{const},$$

where integration goes over the whole space. However, due to the lack of relativistic invariance, the Schrödinger equation gives only an approximate description of the physical system.

On the other hand, (1.45) is relativistically invariant. Let us try to obtain an analogue of the local conservation law in (1.49) for the function f. (We will consider f to be complex even though the electromagnetic field is real.) To this end, write (1.45),

$$\frac{\partial^2 f}{\partial t^2} - \nabla^2 f = 0, \tag{1.50}$$

and, after complex conjugation,

$$\frac{\partial^2 f^*}{\partial t^2} - \nabla^2 f^* = 0. \tag{1.51}$$

Combining (1.50) and (1.51) we arrive at the local conservation law

$$f^* \frac{\partial^2 f}{\partial t^2} - f \frac{\partial^2 f^*}{\partial t^2} = \frac{\partial}{\partial t} \left(f^* \frac{\partial f}{\partial t} - f \frac{\partial f^*}{\partial t} \right) = \mathrm{div}(f^* \boldsymbol{\nabla} f - f \boldsymbol{\nabla} f^*) .$$

Thus we have constructed a local function with a conserved integral, and we may try to interpret it as a probability density:

$$\rho(\mathbf{r}, t) = i \left(f^* \frac{\partial f}{\partial t} - f \frac{\partial f^*}{\partial t} \right) . \qquad (1.52)$$

It is obvious from (1.52) that f has to be complex: if it were real, (1.52) would be identically zero and it would be impossible to use f for the quantum mechanical description of the propagation of free waves.

Unlike the case of non-relativistic quantum mechanics, $\rho(\mathbf{r}, t)$ in (1.52) contains derivatives. This is a consequence of the fact that (1.50) is of second order in time. Hence, to determine the wave function completely, both the function and its derivative have to be fixed at the initial moment. This is the condition on the experiment which determines the initial state of the system. A classical analogy – the electromagnetic potential: $\mathbf{A}$ satisfies the second order equation, but measuring the fields $\mathbf{E}$ and $\mathbf{H}$ fixes both $\mathbf{A}$ and its time derivative, $\dot{\mathbf{A}}$.

Another, more serious problem is that $\rho(\mathbf{r}, t)$, as defined by (1.52), might turn *negative*, and a negative $\rho(\mathbf{r}, t)$ does not admit probabilistic interpretation. To avoid this difficulty, we have to choose only those solutions of (1.50) for which $\rho(\mathbf{r}, t)$ is positive.

First, consider general solutions of (1.50). We look for a solution in the form of a Fourier series

$$f(x) = \sum_k e^{-ikx} \tilde{f}(k) ,$$

where $kx = k_0 x_0 - k_1 k x_1 - k_2 x_2 - k_3 x_3$; k_0, k_1, k_2 and k_3 are arbitrary real numbers. Hence,

$$\Box f(x) = \sum_k (-k_\mu^2) e^{-ikx} \tilde{f}(k) = 0 , \qquad k_\mu^2 = k_\mu k_\mu \equiv k_0^2 - k_1^2 - k_2^2 - k_3^2 .$$
$$(1.53)$$

Equation (1.53) has two obvious solutions

$$k_0 = \pm |\mathbf{k}| .$$

Fixing k_0 to be positive from now on, $k_0 \equiv |\mathbf{k}|$, we can write the general solution as

$$f(x) = f_+(x) + f_-(x) ,$$

where

$$f_+(x) = \sum_{\mathbf{k}} e^{-i(k_0 x_0 - \mathbf{k}\cdot\mathbf{r})} f(\mathbf{k}),$$

$$f_-(x) = \sum_{\mathbf{k}} e^{-i(-k_0 x_0 - \mathbf{k}\cdot\mathbf{r})} f(k) = \sum_{k'} e^{i(k_0 x_0 - \mathbf{k}'\cdot\mathbf{r})} f(k')$$

are the positive-frequency and negative-frequency solutions, respectively. Thus, unlike the case of the Schrödinger equation, here we have two complex solutions with opposite frequencies. Let us show that one of the two, f_+ or f_-, can be taken to represent the photon wave function.

Choose $\Psi = f_+$ as the wave function. As was shown above, the integral of the local density

$$\rho = i\left(f_+^*(x) \frac{\partial f_+(x)}{\partial t} - f_+(x) \frac{\partial f_+^*(x)}{\partial t} \right) \equiv f_+^*(x) \, i\overleftrightarrow{\frac{\partial}{\partial t}} \, f_+(x) \qquad (1.54)$$

is conserved. Indeed,

$$\int \rho \, d^3 r = \sum_{\mathbf{k},\mathbf{k}'} f(k) f^*(k') \, i \int \left[e^{ikx}(-ik_0') e^{-ik'x} - ik_0 e^{-ik'x} e^{ikx} \right] d^3 r$$

$$= (2\pi)^3 \sum_{\mathbf{k}} f(k) f^*(k) \, 2k_0$$

does not depend on t. It is also positive definite. At the same time, for the negative-frequency solution (1.54) the integral $\int \rho \, d^3 r$ is negative. We could easily make it positive by simply changing sign in the expression (1.54) that defines ρ, if we were to choose f_- to describe the photon wave function. We shall stick to the choice $\Psi = f_+$ which is motivated by the analogy with the non-relativistic case, where the wave function depends on time as $\exp(-iEt)$ with $E > 0$.

The next question is whether the function ρ can be interpreted as the *local* probability density. This is possible only if $\rho(\mathbf{r}, t)$ itself, and not simply its integral, is positive. Generally speaking, this requirement is not satisfied, since ρ is a sum of oscillating exponents. The only exception is the stationary case, when

$$f_+(x) = e^{-i\omega t} \, \Psi_\omega(\mathbf{r})$$

and

$$\rho(\mathbf{r}, t) = 2\omega \cdot |\Psi_\omega(\mathbf{r})|^2$$

does not depend on time and is positive definite.

We see that the photon wave function can be chosen as a positive-frequency solution of the d'Alembert equation (1.50). In this case the function $\rho(\mathbf{r}, t)$ can be defined in such a way that its integral over the whole space is positive and conserved in time. For stationary states ρ can be interpreted as the probability density as in non-relativistic quantum mechanics.

The reason why the probabilistic interpretation seems to fail for the non-stationary states is deeply rooted in the nature of relativistic theory. In non-relativistic quantum mechanics the object remains self-identical in the course of measurement. In relativistic theory, as we shall see shortly, the number of particles is not conserved. Localization of the photon in the course of interaction (measurement) inevitably leads to creation of other photons, and the notion of the one-photon wave function becomes meaningless.

We have to establish two more facts: first to find an analogue of the orthogonality condition for the wave functions which correspond to different $\mathbf{k}$, and second to write the normalized photon wave function.

In the non-relativistic case we had

$$\int \Psi_{\mathbf{k}}^*(x)\, \Psi_{\mathbf{k}'}(x)\, d^3r \;=\; (2\pi)^3 \delta(\mathbf{k}' - \mathbf{k})\,. \tag{1.55}$$

The condition (1.55) follows from an equation analogous to the equation of continuity (see e.g. Section 20 in Landau and Lifshitz: *Quantum Mechanics* [1]). In our case the corresponding expression is

$$\int f_{+\mathbf{k}}^*(x)\, i \overleftrightarrow{\frac{\partial}{\partial t}}\, f_{+\mathbf{k}'}(x)\, d^3r \;=\; (2\pi)^3 \delta(\mathbf{k}' - \mathbf{k})\,. \tag{1.56}$$

In this sense the negative- and positive-frequency solutions are always orthogonal, i.e.

$$\int f_{-\mathbf{k}}^*(x)\, i \overleftrightarrow{\frac{\partial}{\partial t}}\, f_{+\mathbf{k}'}(x)\, d^3r \;=\; 0\,.$$

Our free particle will be described by a plane wave

$$f_{+\mathbf{k}}(x) = e^{-ikx}\, f(\mathbf{k})\,, \qquad k_0 = |\mathbf{k}|\,.$$

Substituting this into (1.56), we obtain the normalization condition for the amplitude $f(\mathbf{k})$:

$$f(\mathbf{k})\, f^*(\mathbf{k}')\, 2k_0\, (2\pi)^3 \delta(\mathbf{k} - \mathbf{k}') = (2\pi)^3 \delta(\mathbf{k} - \mathbf{k}')\,,$$

which gives $f(\mathbf{k}) = 1/\sqrt{2k_0}$ and

$$f_{+\mathbf{k}}(x) = \frac{e^{-ikx}}{\sqrt{2k_0}}\,. \tag{1.57}$$

The plane wave $f_{+\mathbf{k}}(x)$ describes the freely propagating photon with momentum $\mathbf{k}$ and energy $k_0 = |\mathbf{k}|$.

In fact, it is not quite photons that we have been discussing so far, since photons are vector particles (i.e. they are described by the four-vector potential A_μ). Let us now repeat our previous considerations for genuine photons. We again separate the positive- and negative-frequency parts of the solutions of the d'Alembert equation for A_μ:

$$\Box A_\mu = 0. \tag{1.58}$$

We start from the real classical potential A_μ and look for a solution in the form

$$A_\mu(x) = \int \frac{d^3k}{(2\pi)^3} \left[a_\mu(\mathbf{k})e^{-ikx} + a_\mu^*(\mathbf{k})e^{ikx} \right]. \tag{1.59}$$

Substituting (1.59) into (1.58), we get, as before,

$$k^2 \equiv k_\mu^2 = 0 \qquad \text{or} \qquad k_0 = \pm|\mathbf{k}|.$$

Hence, the wave function of a photon with $k_0 = |\mathbf{k}|$ can be written as

$$\psi_\mu(x) = \int \frac{d^3k}{(2\pi)^3} a_\mu(\mathbf{k}) e^{-ikx}. \tag{1.60}$$

For a normalized state with momentum $\mathbf{k}$ we find

$$a_\mu(\mathbf{k}) = \frac{e_\mu(\mathbf{k})}{\sqrt{2k_0}}, \tag{1.61}$$

where e_μ is a unit polarization vector. The derivation is the same as for (1.57).

The Lorentz condition leads to

$$\frac{\partial \psi_\mu}{\partial x_\mu} = -i \int \frac{d^3k}{(2\pi)^3} \frac{k_\mu e_\mu(\mathbf{k})}{2k_0} e^{-ikx} = 0,$$

i.e.

$$k_\mu e_\mu(\mathbf{k}) = 0. \tag{1.62}$$

This is the condition for the four-dimensional transversality of the photon. What does it mean? Generally speaking, one can introduce four independent unit vectors e_μ^λ in four-dimensional space. Due to (1.62) there remain three independent vectors, orthogonal to k_μ. However, since $k_\mu^2 = 0$, one of these vectors will be proportional to the four-vector k_μ which is 'orthogonal to itself'. In other words, in Minkowski space (as opposed to the case of Euclidean metrics) it is impossible to construct three independent vectors which are orthogonal to a *light-like* vector and differ from it.

Indeed, let us choose the reference frame so that

$$\mathbf{k} \parallel \mathbf{z}, \quad \text{i.e.} \quad k_\mu = (k_0, 0, 0, k_z), \quad k_0 = k_z.$$

Two vectors, orthogonal to k_μ, are

$$
\begin{aligned}
e_\mu^{(1)} &= (0, 1, 0, 0) \equiv (e_0, e_x, e_y, e_z), \\
e_\mu^{(2)} &= (0, 0, 1, 0),
\end{aligned}
$$

while the third vector $e_\mu^{(3)}$ is parallel to k_μ. (Indeed, given $k_0 = k_z$, from $k_0 e_0^{(3)} - k_z e_z^{(3)} = 0$ immediately follows $e_0^{(3)} = e_z^{(3)}$.) Consequently, both vectors $e_\mu^{(3)}$ and k_μ have the form $(a, 0, 0, a)$, i.e. they differ only by a numerical factor. This third polarization (the so-called *longitudinal* polarization) does not count, however, as a degree of freedom of the photon. The term in the potential corresponding to $a_\mu^{(3)}(\mathbf{k}) = e_\mu^{(3)}(\mathbf{k})/\sqrt{2k_0} \propto k_\mu$ does not enter the gauge independent electromagnetic field strengths, $\mathbf{E}$ and $\mathbf{H}$.

For a real photon, we can always get rid of the polarization $e_\mu^{(3)} \propto k_\mu$ with the help of a gauge transformation, and therefore its existence cannot affect any physical results. In reality, the Maxwell equations are invariant under the gauge transformation of potentials

$$A_\mu \to A_\mu + \frac{\partial f}{\partial x_\mu}. \qquad (1.63)$$

This transformation does not violate the Lorentz condition, provided $\Box f = 0$. Introducing the Fourier representation of $f(x)$,

$$f(\mathbf{k}) = \int f(x)\, e^{ikx}\, d^3r,$$

gives

$$\frac{\partial f(x)}{\partial x_\mu} = \int \frac{d^3k}{(2\pi)^3} e^{-ikx}\, (-ik_\mu)\, f(\mathbf{k}).$$

This means that in momentum space the gauge transformation (1.63) leads to the transformation of the photon wave function

$$a_\mu(\mathbf{k}) \to a_\mu(\mathbf{k}) - ik_\mu f(\mathbf{k}).$$

Since the gauge-dependent addition to the potential corresponds to $e_\mu^{(3)}$, the contribution of the longitudinal polarization can always be turned into zero by the proper choice of $f(\mathbf{k})$.

Hence, the real photon has only two independent transverse polarizations. Although the photon spin equals one, only two of its spin states (with projections of spin plus or minus one on the direction of motion) can contribute to physical observables. This is a consequence of the photon having no mass ($k_\mu^2 = m^2 = 0$).

So, the photon wave function can be written as a sum of the two polarization contributions,

$$\psi_\mu = \sum_{\lambda=1,2} \int \frac{d^3 k}{(2\pi)^3} \frac{e_\mu^\lambda(\mathbf{k})}{\sqrt{2k_0}} e^{-ikx} \, C(k,\lambda) \equiv \sum_{\lambda=1,2} \int \frac{d^3 k}{(2\pi)^3} \psi_\mu^{\lambda \mathbf{k}}(x) C(k,\lambda),$$

with $\quad \psi_\mu^{\lambda \mathbf{k}}(x) = \dfrac{e_\mu^\lambda(\mathbf{k})}{\sqrt{2k_0}} e^{-ikx}.$

We come to the conclusion that the two *spatial* components of the vector potential A_μ should play the rôle of the wave functions of transversally polarized real photon states. In order to construct the photon probability density in analogy with (1.54), we have to sum the product of the wave functions, $\psi_\mu^* \psi_\mu$, over the vector index μ. Given our agreement about the Lorentz space metric, $e_\mu^{(1)} e_\mu^{(1)} = e_\mu^{(2)} e_\mu^{(2)} = -1$, this implies contracting the two vectors with the *minus* sign in order to preserve positivity of ρ for physical spatial polarization states:

$$\rho = (-g_{\mu\nu}) \cdot \psi_\mu^*(x) \, i \frac{\overleftrightarrow{\partial}}{\partial t} \, \psi_\nu(x).$$

We write the normalization condition for the photon wave functions accordingly:

$$-g_{\mu\nu} \int \left(\psi_\mu^{\lambda_2 k}(x) \right)^* i \frac{\overleftrightarrow{\partial}}{\partial t} \, \psi_\nu^{\lambda_1 k'}(x) \, d^3 r \tag{1.64}$$
$$= - e_\mu^{\lambda_2 *} e_\mu^{\lambda_1} (2\pi)^3 \delta(\mathbf{k} - \mathbf{k'}) = \delta_{\lambda_1 \lambda_2} (2\pi)^3 \delta(\mathbf{k} - \mathbf{k'}).$$

The photon wave function we have thus constructed has a simple classical interpretation. Writing the vector potential in the form (1.59) and calculating classically the mean energy of the electromagnetic field,

$$\int_{V=1} \frac{\mathbf{E}^2 + \mathbf{H}^2}{2} \, d^3 r = \omega,$$

we conclude that the normalization of our amplitude a_μ corresponds to having exactly one photon in a unit volume.

1.5 Photons in an 'external field'

1.5.1 Relativistic propagator

Let us try to find the propagator for a relativistic particle which is described by the positive-frequency wave function $f_+(x)$. (For the time being, we shall again suppress the indices.)

In non-relativistic theory, we have obtained the propagator in the form of a sum over all eigenfunctions,

$$K(x_2, x_1) = \sum_n \Psi_n(x_2) \, \Psi_n^*(x_1), \qquad (1.65)$$

in such a way that the evolution of the quantum system is described by

$$\Psi(x_2) = \int K(x_2, x_1) \, \Psi(x_1) \, d^3r_1, \qquad x_i = (t_i, \mathbf{r_i}). \qquad (1.66)$$

In relativistic theory there are two additional considerations:

(1) The time-derivative of the wave function should be included along with Ψ in the propagation law, since together they determine the initial state.

(2) Since we have chosen the positive-frequency solution of the d'Alembert equation to represent the wave function, we should take care not to generate *negative* frequencies when the propagator is acting on the initial wave function $\Psi(x_1)$.

Let us try to write the relativistic propagator in a form analogous to (1.65),

$$K(x_2, x_1) = \sum_n f_n^+(x_2) \, f_n^{+*}(x_1), \qquad (1.67)$$

and alter the propagation law (1.66) in the following way:

$$f^+(x_2) = \int K(x_2, x_1) \, i \frac{\overleftrightarrow{\partial}}{\partial t_1} \, f^+(x_1) \, d^3r_1. \qquad (1.68)$$

It is easy to check that equations (1.67) and (1.68) properly describe the evolution of the wave function of a relativistic particle. Since any wave function can be expanded as a superposition of stationary states f_n^+, it suffices to check the propagation of a single stationary state. Taking into account the relativistic orthogonality condition (1.56), for $f^+(x_1) = f_m^+(x_1)$ we obtain

$$f^+(x_2) = \sum_n f_n^+(x_2) \int f_n^{+*}(x_1) \, i \frac{\overleftrightarrow{\partial}}{\partial t_1} \, f_m^+(x_1) \, d^3r_1 = f_m^+(x_2). \qquad (1.69)$$

In the non-relativistic case we had a relation analogous to (1.69) with the propagator (1.65) and an orthogonality condition (1.55).

The fact that (1.68) contains the time-derivative is in accord with the d'Alembert equation being of second order in time: it provides the solution with given initial conditions, i.e. with given values of the function and its time-derivative at the initial moment of time. We conclude that (1.67) properly evolves the photon wave function, that is an arbitrary positive-frequency solution of the d'Alembert equation. At the same time, it does not propagate (and does not generate) any negative-frequency states.

Let us calculate the propagator of a free relativistic massless particle. Since we already know the normalized wave function

$$f_n = \frac{e^{-ikx}}{\sqrt{2k_0}},$$

inserting it into (1.67) immediately gives us the propagator:

$$K(x_2, x_1) = \int \frac{d^3k}{(2\pi)^3} \frac{e^{-ik(x_2-x_1)}}{2k_0}, \qquad k_0 = |\mathbf{k}|. \qquad (1.70)$$

Our propagator is relativistically invariant. To see it explicitly we use the relation

$$\delta(f(x)) = \sum_i \frac{1}{|f'(x_i)|}\delta(x - x_i); \qquad f(x_i) = 0$$

to write

$$K(x_2, x_1) = \int \frac{d^4k}{(2\pi)^3} \delta(k^2) e^{-ik(x_2-x_1)} \theta(k_0). \qquad (1.71)$$

(The step function $\theta(k_0)$ is inserted in the integrand to avoid the unwanted solution $k_0 = -|\mathbf{k}|$.) The four-dimensional integration d^4k, the variables k_μ^2 and $k_\mu(x_2 - x_1)_\mu$ entering (1.71) are manifestly Lorentz invariant. So is the *sign* of the energy: the sign of the zero-component of k_μ does not depend on the reference frame for light-like (and time-like) four-vectors, $k^2 \geq 0$.

Thus, we have managed to construct a relativistically invariant propagator $K(x_2, x_1)$ in spite of having restricted ourselves to positive frequencies only. It is easy to see that $K(x_2, x_1)$ satisfies

$$\Box_2 K(x_2, x_1) = 0 \qquad (1.72)$$

(where $\Box_2$ means the d'Alembertian in x_2): substituting (1.71) into (1.72), we get

$$\int d^4k \, k^2 \delta(k^2) \ldots = 0.$$

1.5.2 Relativistic interaction

We now try to introduce an interaction $V(x)$ into our relativistic picture. Similarly to the non-relativistic case, we would like to write something within the logic of 'free propagation – point-like interaction – free propagation',

$$\int K(x_2 - x)\, V(x)\, K(x - x_1)\, d^4x \,, \qquad (1.73)$$

corresponding to the graph

$$x_1 \qquad\qquad x \qquad\qquad x_2$$

In the non-relativistic case we have added to this expression the condition $t_1 < t < t_2$ to ensure that the particle was first created and only afterwards did it interact. This was achieved by introducing the Green function

$$G(x_2 - x_1) \;=\; \theta(t_2 - t_1)\, K(x_2 - x_1)\,.$$

One might wonder if imposing such a condition is possible in relativistic theory, in which the amplitude (1.73) has to be relativistically invariant. Generally speaking, the time-ordering condition $t_1 < t$ is not relativistically invariant. It is invariant only for time-like intervals $(x - x_1)^2 > 0$, in which case the time sequence of events does not depend on the reference frame. If $K(x - x_1)$ were different from zero only for $(x - x_1)^2 > 0$, we could impose such a condition. However, $K(x - x_1)$ does not vanish for space-like intervals $(x - x_1)^2 < 0$ (see (1.71)), and therefore $t > t_1$ makes no sense: insisting on $t_1 < t < t_2$ would lead to a non-invariant expression for the transition amplitude.

Our reasoning can be checked directly. Let us write

$$\tilde{G} = \theta(t_2 - t_1)\, K(x_2 - x_1)$$

and see what equation this function will satisfy. Acting with the d'Alembert operator on $\tilde{G}$, we obtain

$$\left(\frac{\partial^2}{\partial t_2^2} - \boldsymbol{\nabla}_2^2 \right) \tilde{G}(x) = \theta(t_2 - t_1) \left[\frac{\partial^2 K}{\partial t_2^2} - \boldsymbol{\nabla}_2^2 K \right] + \frac{\partial \theta}{\partial t_2} \frac{\partial K}{\partial t_2} + \frac{\partial^2 \theta}{\partial t_2^2} K \,.$$

In this expression the first term in the right-hand side is zero because of (1.72), while the other terms are obviously not relativistically invariant. Thus the condition $t_1 < t < t_2$ is incompatible with Lorentz invariance and is therefore meaningless. In principle, we could try to look for some other function $\tilde{K}$ such that the product $\theta(t_2 - t_1) \cdot K(x_2 - x_1)$ would satisfy a relativistically invariant equation. The 'new propagator' K, however, would inevitably generate negative-frequency states.

Hence, we have failed to reconcile two conditions:

(1) The propagator should contain only positive frequencies (because only in this case is the probabilistic interpretation possible).

(2) The interaction should take place at time t between t_1 and t_2 (i.e. the requirement of causality).

Which of the two to sacrifice? Giving up the first requirement would mean abandoning the probability interpretation of quantum mechanics. For this reason, we had better look for a way to reconsider causality as it is formulated in the second condition. Could not, for example, the interaction occur *before* time t_1 when the particle was born, as shown by the graph?

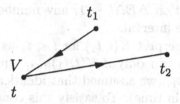

(Hereafter we will assume that the time axis is directed from left to right.)

Indeed, the interpretation of this graph can be changed. Let us say that it corresponds to the creation of two particles at time t, one of which disappears at time t_1. In this case the causality remains valid, and diagrams of this type are meaningful.

The new interpretation we are looking for becomes even more transparent in the case of two interactions,

If $t' < t$, this diagram can be rearranged similarly to the previous one:

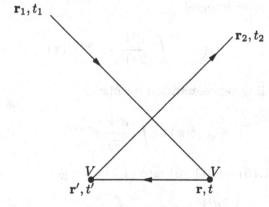

Now we are ready to interpret it: at time t' two particles were created, at t a particle propagating from r_1 and one propagating from r' annihilate, and at t_2 there remains only one particle. (Here the particle we detect at t_2 is actually not *the* particle that was created at t_1 but, then again, how are we to distinguish them?)

Such an interpretation is possible if we assume that the propagator, in the presence of interaction, includes the processes in which several particles can be present simultaneously. Hence, in order to be able to introduce the interaction in relativistic theory, one has to abandon the idea of conservation of the number of particles. This goes in line with the fact that we have not succeeded in introducing a positively definite local probability density (recall the discussion, page 22). The non-conservation of the number of particles, i.e. the possibility of their production and annihilation, does not contradict any fundamental principles since, due to the uncertainty relation $\Delta E \Delta t \sim 1$, any number of particles can be created for a short time interval.

Obviously, we can interpret $K(t, t_1)$ for $t < t_1$ as a function propagating the particle from t_1 to t only if $K(t, t_1)|_{t<t_1}$ equals $K(t, t_1)|_{t>t_1}$ with the same value of $|t - t_1|$ (we assumed that identical particles propagate forward and backward in time). To satisfy this condition the propagator has to have a discontinuity at $t = t_1$. As a result, this function will no longer satisfy the homogeneous d'Alembert equation (1.72), but will turn out to be its Green function.

1.5.3 Relativistic Green function

We need to find the Green function of the d'Alembert equation,

$$\Box G(x) = -i\delta(x) , \qquad (1.74)$$

and establish its connection with the propagator (1.71). Let us represent $G(x)$ as the Fourier integral

$$G(x) = \int \frac{d^4 k}{(2\pi)^4 i} e^{-ikx} G(k) . \qquad (1.75)$$

The corresponding representation for $\delta(x)$ is

$$\delta(x) = \int \frac{d^4 k}{(2\pi)^4} e^{-ikx} . \qquad (1.76)$$

Substituting (1.76) and (1.75) into (1.74), we get

$$\Box G(x) = \int \frac{d^4 k}{(2\pi)^4 i} (-k^2) \cdot e^{-ikx} G(k) = -i \int \frac{d^4 k}{(2\pi)^4} e^{-ikx} ,$$

which leads to

$$G(k) = -\frac{1}{k^2} . \qquad (1.77)$$

Thus,

$$G(x) = -\int \frac{d^4k}{(2\pi)^4 i} \, e^{-ikx} \, \frac{1}{k^2}. \qquad (1.78)$$

The integrand in (1.78) has two poles in k_0, namely, $k_0 = \pm|\mathbf{k}|$. To make the integral (1.78) well defined, the poles should be shifted from the real axis. There are four possibilities for shifting the poles from the real axis into the complex plane, shown in Fig. 1.1.

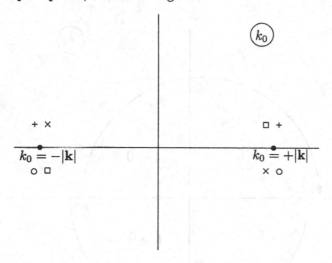

Fig. 1.1

We will consider only the two possible configurations of poles and two respective Green functions that are especially important to the theory.

(1) Both poles are in the lower half-plane (marked as ∘). In this case, if $t < 0$, the contour has to be closed in the upper half-plane, and

$$G_{\mathrm{R}} = 0, \qquad t < 0.$$

If $t > 0$, the contour has to be closed in the lower half-plane, and we have

$$G_{\mathrm{R}} = \int \frac{d^3k}{(2\pi)^3} \frac{e^{-i|k|t + i\mathbf{k}\cdot\mathbf{r}}}{2|\mathbf{k}|} - \int \frac{d^3k}{(2\pi)^3} \frac{e^{i|k|t + i\mathbf{k}\cdot\mathbf{r}}}{2|\mathbf{k}|}, \qquad t > 0.$$

The *retarded* Green function G_{R} contains negative frequencies and therefore does not suit us: we cannot use it to describe propagation of relativistic particles.

(2) The pole on the negative axis is shifted upwards, while that on the positive axis is shifted downwards (marked as ×). If $t > 0$, the

contour is closed in the lower half-plane (see Fig. 1.2), and we obtain

$$G = \int \frac{d^3k}{(2\pi)^3} \frac{e^{-i|k|t+i\mathbf{k}\cdot\mathbf{r}}}{2|\mathbf{k}|}, \qquad t > 0, \qquad (1.79)$$

while for $t < 0$ the corresponding expression is

$$G = \int \frac{d^3k}{(2\pi)^3} \frac{e^{i|k|t+i\mathbf{k}\cdot\mathbf{r}}}{2|\mathbf{k}|}. \qquad (1.80)$$

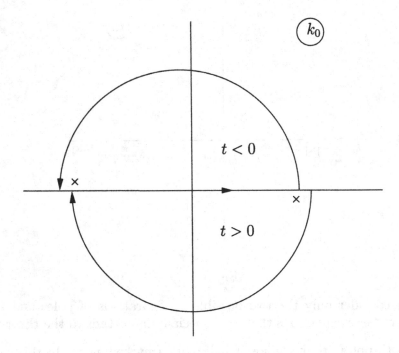

Fig. 1.2

Comparing (1.79) with (1.70), we discover a remarkable fact: for $t > 0$ our Green function coincides with the relativistic propagator (1.71), while for $t < 0$ it contains negative (and only negative) frequencies. Moreover, the phase in the exponent is always negative ($-i|\mathbf{k}|t$ for $t > 0$ and $+i|\mathbf{k}|t$, $t > 0$), and

$$G(x) = G(-x), \qquad -x \equiv (-\mathbf{r}, -t). \qquad (1.81)$$

We may say that G propagates positive (and only positive) frequencies *forward* in time, and negative (and only negative) frequencies *backward* in time. This Green function is called the *causal* or Feynman Green function. As we shall see below, it is this function which describes propagation of relativistic particles in a truly causal manner.

Position of the poles in the Feynman Green function can be described in terms of the substitution $k^2 \to k^2 + i\varepsilon$ in the denominator in (1.78).

1.5.4 Propagation of vector photons

What will change in the previous analysis when we take *spin* into account? The wave function of the photon is a vector $f_\mu^\lambda(x)$. In this case the propagator will depend on the spins of the initial and final states and become a Lorentz tensor, $K = K_{\mu\nu}$. It is straightforward to verify that

$$K_{\mu\nu}(x_2, x_1) = \sum_n \sum_{\lambda=1,2} f_\mu^{n\lambda}(x_2) \left(f_\nu^{n\lambda}(x_1) \right)^* \qquad (1.82)$$

properly propagates stationary photon states $f_\nu^{m\lambda}$. The equation for the photon's *Green function* $D_{\mu\nu}$ will change accordingly:

$$\Box D_{\mu\nu}(x) = i\, g_{\mu\nu} \delta(x) \,, \qquad (1.83)$$

where we have chosen the sign of the right-hand side such that for the physical – spatial – components ($g_{ii} = -1$) it coincides with that in (1.74). Thus,

$$D_{\mu\nu}(x) = g_{\mu\nu} \int \frac{d^4 k}{(2\pi)^4 i} \frac{e^{-ikx}}{k^2 + i\varepsilon} \,. \qquad (1.84)$$

The infinitesimal imaginary part $i\varepsilon$ shifts the poles in the right direction. Indeed, $k^2 + i\varepsilon = 0 \;\rightarrow\; k_0 = \pm\sqrt{\mathbf{k}^2 - i\varepsilon} \rightarrow \pm|\mathbf{k}| \mp i\varepsilon$. For $t > 0$ (1.84) gives

$$D_{\mu\nu}(x) = -g_{\mu\nu} \int \frac{d^3 k}{(2\pi)^3} \frac{e^{-ikx}}{2|\mathbf{k}|} \,, \qquad t > 0\,, \quad k_0 = |\mathbf{k}| \,. \qquad (1.85)$$

This function does not completely coincide with the propagator (1.82), since in

$$K_{\mu\nu}(x) = \int \frac{d^3 k}{(2\pi)^3} \frac{e^{-ikx}}{2|\mathbf{k}|} \sum_{\lambda=1,2} e_\mu^\lambda e_\nu^{\lambda*}$$

the summation goes only over the two physical photon polarizations, while the metric tensor $g_{\mu\nu}$ in (1.85) 'propagates' all *four* independent polarization vectors,

$$-g_{\mu\nu} = \sum_{\lambda=0}^{3} e_\mu^\lambda e_\nu^{\lambda*} \,. \qquad (1.86)$$

This means that (1.85) misses the fact that there are only two independent polarizations. We must correct the Green function. On the other hand, $D_{\mu\nu}$ is the only solution of (1.83), and it seems impossible to alter anything on the right-hand side of the equation without losing relativistic invariance.

What is the nature of the problem? We wanted to have two polarizations in $K_{\mu\nu}$, while in $D_{\mu\nu}$ there are four. This means that we have come to a contradiction with gauge invariance and/or the Lorentz condition. There is nothing strange, however: for an arbitrary interaction potential neither gauge invariance nor the Lorentz condition has to hold. To preserve them, a certain condition should be imposed on the interaction! In classical theory the situation was the same. A solution of the Maxwell equation,

$$\frac{\partial F_{\nu\mu}}{\partial x_\nu} = j_\mu \,,$$

cannot be found for an arbitrary j_μ. It exists only for the *conserved* current,

$$\frac{\partial j_\mu}{\partial x_\mu} = 0 \,.$$

Current conservation follows from the antisymmetry of $F_{\mu\nu}$:

$$\frac{\partial^2 F_{\mu\nu}}{\partial x_\nu \partial x_\mu} = \frac{\partial j_\mu}{\partial x_\mu} = 0 \,.$$

Let us return to comparison of the propagator and the Green function. The propagator was constructed for real photons $k^2 = 0$, while the Green function contains four integrations over k_0 and $\mathbf{k}$, so that the virtual photon is, generally speaking, 'massive': $k^2 \neq 0$. For $k^2 \neq 0$, the Lorentz condition $k_\mu e_\mu = 0$ determines not two but *three* independent vectors which do not coincide with k_μ itself, and we can include all of them in the sum over polarizations in the Green function. These polarizations are:

$$\sum_{\lambda=1}^{2} e_\mu^\lambda e_\nu^{\lambda *} = -g_{\mu\nu}^\perp \,,$$

$$e_\mu^{(3)} e_\nu^{(3)*} = \frac{\left[(k\tau)k_\mu - k^2\tau_\mu\right]\left[(k\tau)k_\nu - k^2\tau_\nu\right]}{k^2\left[(k\tau)^2 - k^2\tau^2\right]}$$

$$= \frac{\left[k_\mu - \frac{k^2}{k_0}\tau_\mu\right]\left[k_\nu - \frac{k^2}{k_0}\tau_\nu\right]}{k^2} \frac{k_0^2}{|\mathbf{k}|^2} \,,$$

where $g_{\mu\nu}^\perp$ is the unit tensor in the (x, y) plane orthogonal to $\mathbf{k} = (0, 0, k_z)$, and τ is the unit time-vector, $\tau = (1; 0, 0, 0)$. Together with the 'scalar' polarization contribution,

$$e_\mu^{(0)} e_\mu^{(0)*} = -\frac{k_\mu k_\nu}{k^2} \,,$$

they form the full metric tensor in (1.86).

Thus, the only way we can improve our Green function (1.84) is to include the sum over three polarizations instead of two. It may be cast in a relativistically invariant form,

$$\sum_{\lambda=1}^{3} e_\mu^\lambda e_\nu^{\lambda*} = -g_{\mu\nu} + \frac{k_\mu k_\nu}{k^2}, \qquad (1.87)$$

as the sum over all four polarization states minus the contribution of the scalar polarization $e_\mu^{(0)}$ parallel to k_μ.

Such an improvement may look rather dangerous, because our new Green function,

$$\tilde{D}_{\mu\nu}(x) = \int \frac{d^4 k}{(2\pi)^4 i} \frac{e^{-ikx}}{-k^2 - i\varepsilon} \left[\frac{k_\mu k_\nu}{k^2} - g_{\mu\nu} \right], \qquad (1.88)$$

has acquired an additional singularity at $k^2 = 0$. This should not worry us too much. To prevent production of *longitudinal* real photons we have, in any case, to organize the interaction in such a way that the terms in the Green function that are proportional to k_μ, k_ν would not contribute to the observables. As a result, the singular term $\propto k_\mu k_\nu / k^2$ in (1.88) can be dropped altogether.

By doing so we would return to the sum over all four polarizations, as in (1.84). This, however, can be tolerated now: imposing the condition on the interaction (current conservation condition) suppresses the production of both scalar and longitudinal photon states, so that only two physical polarizations in $g_{\mu\nu}$ in (1.84) give non-vanishing contributions on the mass shell. Indeed, the contribution of 'superfluous' polarizations is

$$e_\mu^{(3)} e_\nu^{(3)*} + e_\mu^{(0)} e_\nu^{(0)*} = \frac{\tau_\mu \tau_\nu}{|\mathbf{k}|^2} \cdot k^2 + \text{ terms proportional to } k_\mu \text{ and/or } k_\nu.$$

The terms $\propto k_\mu, k_\nu$ do not contribute due to conservation of current (the condition we will have to impose on the interaction), while the remaining term vanishes for $k^2 = 0$. It is responsible for the instantaneous Coulomb interaction between charged particles, which is described in our language by an exchange of *virtual* longitudinally polarized photons.

It is easy to see that, whether we choose to sum over four, as in (1.84), or over three polarizations, see (1.88), the photon Green function satisfies a symmetry relation similar to (1.81):

$$D_{\mu\nu}(x) = D_{\nu\mu}(-x). \qquad (1.89)$$

This means that $D_{\mu\nu}$ describes not only the process

$$t_2 > t_1$$

$$\underset{x_1}{\xrightarrow{\hspace{4cm}}} \hspace{1cm} x_2$$

(the creation of a photon at x_1 and its disappearance at x_2), but also a process which goes back in time:

$$t_2 < t_1$$

x_2 $\qquad\qquad\qquad\qquad\qquad\qquad\qquad\qquad\qquad$ x_1

The latter can be interpreted as the creation of a particle at the point x_2 and its propagation to x_1 and, according to (1.89), *this particle is identical to a photon.* Examples of the processes with different time ordering were given by diagrams on page 29.

Let us summarize what we have obtained so far. We have constructed the photon wave function

$$\psi_\mu = \frac{e_\mu^\lambda e^{-ikx}}{\sqrt{2|\mathbf{k}|}}$$

and the Green function

$$D_{\mu\nu}(x) = g_{\mu\nu} \int \frac{d^4k}{(2\pi)^4 i} \frac{e^{-ikx}}{k^2 + i\varepsilon}.$$

In addition, the wave function

$$\Psi(x) = e^{-ipx}, \quad p_0 = \frac{\mathbf{p}^2}{2m},$$

and the Green function

$$G(x) = \int \frac{d^4p}{(2\pi)^4 i} \frac{e^{-ipx}}{\frac{\mathbf{p}^2}{2m} - p_0 - i\varepsilon}$$

of a non-relativistic particle are known. This is already sufficient for the construction of quantum electrodynamics (QED) of non-relativistic particles, equivalent to the usual quantum theory of radiation. It makes more sense, however, to construct the electrodynamics of *relativistic* particles, which in the non-relativistic limit reproduces the non-relativistic results. For that purpose, let us consider massive relativistic particles.

1.6 Free massive relativistic particles

A free particle at rest can be characterized by two additive integrals of motion: the energy, which is equal to the mass m of the particle, and the internal angular momentum – the spin J. We shall classify all particles by their masses and spins. We have to construct a theory for free particles of arbitrary masses and spins, moving with any velocities, i.e. having arbitrary momenta $\mathbf{p}$.

In order to understand which features of the theory are connected with relativity and which ones with spin, let us begin with spin zero particles $J = 0$; as is well known, there are many particles of this type, e.g. the pions (π^-, π^0, π^+), with mass $m \simeq 140$ MeV.

Classical relativistic invariance leads to the relation

$$E^2 = \mathbf{p}^2 + m^2 \tag{1.90}$$

between energy E and momentum $\mathbf{p}$. Correspondingly, a quantum mechanical particle with momentum $\mathbf{p}$ is described by the wave function

$$\Psi(x) \sim e^{i\mathbf{p}\cdot\mathbf{r} - iEt}, \tag{1.91}$$

where E and $\mathbf{p}$ are related by (1.90).

Let us find an equation for the wave function in (1.91). Consider

$$\left(\frac{\partial^2}{\partial t^2} - \nabla^2 + m^2 \right) \Psi(x) = 0. \tag{1.92}$$

Substituting (1.91) into (1.92), we get (1.90), i.e. the classical equation (1.90) corresponds to the quantum mechanical equation (1.92). Trying to introduce the probability density $\rho(x)$, we face the same difficulties as in the case of the electromagnetic field:

$$\rho(x) \neq |\Psi(x)|^2, \quad \text{and} \quad \int |\Psi(x)|^2 \, d^3r \neq \text{const.}$$

Again, $\Psi^* i \overset{\leftrightarrow}{\frac{\partial}{\partial t}} \Psi$ which satisfies the equation of continuity

$$\frac{\partial}{\partial t} \left[\Psi^*(x) i \overset{\leftrightarrow}{\frac{\partial}{\partial t}} \Psi(x) \right] = \text{div } i \left[\Psi^*(x) \nabla \Psi(x) - \Psi(x) \nabla \Psi^*(x) \right] \tag{1.93}$$

is conserved, and

$$\rho(x) = \Psi_+^*(x) \, i \overset{\leftrightarrow}{\frac{\partial}{\partial t}} \, \Psi_+(x) \tag{1.94}$$

is the only possible expression that we can take for the probability density. Here Ψ_+ is the positive-frequency solution of (1.92) corresponding to $E = \sqrt{\mathbf{p}^2 + m^2}$, and Ψ_- is the negative-frequency solution, which corresponds to $E = -\sqrt{\mathbf{p}^2 + m^2}$.

Henceforth, we shall write $\Psi_+ = \Psi$. In the case of the stationary state $\Psi(x) = \Psi(\mathbf{r}) e^{-iEt}$, the function $\rho(\mathbf{r})$ plays the rôle of the probability density

$$\rho(\mathbf{r}) = 2E \, |\Psi(\mathbf{r})|^2$$

in the same way as for the photon. Similarly, we can introduce the wave function

$$\Psi(x) = \frac{e^{-ipx}}{\sqrt{2p_0}}, \quad p_0 = \sqrt{\mathbf{p}^2 + m^2} \tag{1.95}$$

and the Green function

$$G(x) = \int \frac{d^4p}{(2\pi)^4 i} \frac{e^{-ipx}}{m^2 - p^2 - i\varepsilon} \tag{1.96}$$

for a free particle. (In fact, in the integral representation (1.84) for $D_{\mu\nu}(x)$ the squared momentum enters with the same negative sign as in the integral representation (1.96) for $G(x)$. This is because only the spatial components of $g_{\mu\nu}$ are effective for the photon, and $g_{11} = g_{22} = -1$.) As before,

$$G(x) = G(-x).$$

It also follows from (1.96) that for $t > 0$

$$G(x) = \int \frac{d^3p}{(2\pi)^3} \frac{e^{-i\sqrt{\mathbf{p}^2+m^2}\,t+i\mathbf{p}\cdot\mathbf{r}}}{2\sqrt{\mathbf{p}^2 + m^2}},$$

i.e. positive frequencies propagate forward in time.

1.7 Interactions of spinless particles

How can an interaction be described in relativistic quantum theory? There is no potential, there are no forces – all these are entirely non-relativistic notions. Moreover, the field is also represented by particles ('quantized'). Thus, we have nothing but various particles characterized by their masses and spins.

Let us consider two spinless particles with different masses m_1 and m_2; their wave functions are Ψ_1, Ψ_2. The Green function of the first particle is represented by a solid line, that of the second one by a dotted line:

$$\overline{}_{x_1 \qquad\qquad\qquad\qquad\qquad x_2} = G_1(x_2 - x_1),$$

$$\cdots\cdots\cdots\cdots\cdots\cdots\cdots\cdots_{x_3 \qquad\qquad\qquad\qquad\qquad x_4} = G_2(x_4 - x_3).$$

If these two particles interact, what will happen? If we assume that there are no other particles around, for point-like objects there is only one possibility: they collide and scatter at a point x, as is shown in Fig. 1.3.

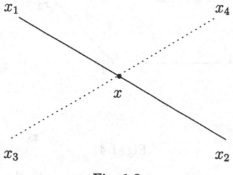

Fig. 1.3

How do we write the amplitude for this process? Following the picture (two free particles propagate from x_1, x_3 to the point x, interact, then propagate to x_2, x_4) let us write

$$G_{12}(x_2, x_4; x_1, x_3) \tag{1.97}$$
$$= \int G_1(x_2-x)G_2(x_4-x)V(x)G_1(x-x_1)G_2(x-x_3)d^4x,$$

where $V(x)$ is an interaction amplitude. This amplitude cannot depend on any relative coordinates because we have assumed that the interaction is local. Moreover, due to homogeneity of space-time, it cannot depend on the position of the point where the particles meet either. Consequently, $V(x) = \text{const} = \lambda$. Thus, we obtained a definite expression for the transition amplitude which contains only one overall unknown constant factor.

The integration in (1.97) goes over all points x in space-time. In the region

$$t_1, t_3 < t < t_2, t_4$$

the interpretation is clear. We know, however, that it is impossible to maintain a restriction on the region of integration since in the relativistic case $G(\Delta t)$ does not vanish for $\Delta t < 0$. Other regions will contribute as well, e.g.

$$t_1 < t < t_2, t_3, t_4.$$

In the latter case our previous interpretation of the process does not make sense literally, and it is natural to redraw the space-time diagram as shown in Fig. 1.4.

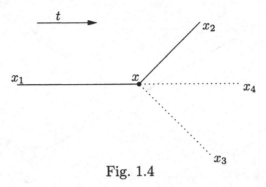

Fig. 1.4

This means that a particle propagates from x_1 to the point x where it decays into three particles as result of the interaction. In other words, due to relativistic invariance, the process described by Fig. 1.3 can only be considered together with particle creation processes like Fig. 1.4. The reason is that the integral (1.97) necessarily contains an integration region where the process can be interpreted only if we accept particle non-conservation. But this is exactly what happens in nature! As we have discussed in Section 1.5.2, non-conservation of the number of particles is a highly non-trivial fundamental consequence of relativistic theory which is confirmed by experiment.

Let us look at the particle production process in Fig. 1.4 and try to write its amplitude according to our rules (propagation of particle 1 from x_1 to x, point-like decay, propagation of particle 1 and two particles of type 2 from x to their final destination points x_2, x_3 and x_4):

$$G(x_2, x_3, x_4; x_1) \tag{1.98}$$
$$= \int G_1(x_2 - x) G_2(x_3 - x) G_2(x_4 - x) \tilde{V}(x) G_1(x - x_1) d^4x,$$

where $\tilde{V}(x) = \tilde{\lambda}$ is a space- and time-independent decay amplitude. On the other hand, as we already know, this process is also contained in (1.97), so that for the same values of the coordinates x_i (1.97) and (1.98) must coincide. Comparing them we observe that all the elements of the two expressions match except the Green functions G_2 connecting the points x and x_3. However, these two elements are also equal due to the symmetry $G_2(x_3 - x) = G_2(x - x_3)$, and we derive

$$\lambda = \tilde{\lambda}.$$

We see that the expression for the amplitude of a *scattering* process in a region with 'strange' time ordering coincides with what we could have expected for a completely different process – that of *particle creation*. The very existence of a scattering process represented by Fig. 1.3 implies the existence of a number of other processes. Among them are the 'decay' process like the one in Fig. 1.4, or the process of annihilation of two particles of the first type into two particles of the second type (Fig. 1.5).

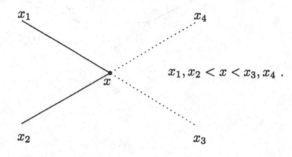

$$x_1, x_2 < x < x_3, x_4 \ .$$

Fig. 1.5

Thus, we conclude:

(1) Due to the symmetry of the Green function, processes of particle 'transmutation' and 'decays' arise automatically from the 'scattering' process shown in Fig. 1.3.

(2) The amplitudes λ of all these processes are equal, i.e. we derive from the requirement of Lorentz invariance not only existence of different processes, but also the connection between them.

(3) For consistent interpretation of such processes it is necessary to assume that the function $G(x - x_1)$ describes propagation of a particle from x_1 to x, when $t_1 < t$, and from x to x_1, when $t_1 > t$.

Let us see what will happen if the particles interact more than once as shown in Fig. 1.6.

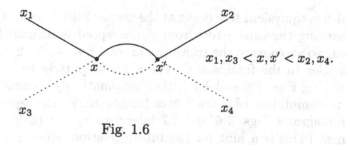

$$x_1, x_3 < x, x' < x_2, x_4.$$

Fig. 1.6

Is this all? In Fig. 1.6 we supposed that it was the particles coming from x_1 and x_3 that met at x. An alternative double interaction picture is shown in Fig. 1.7.

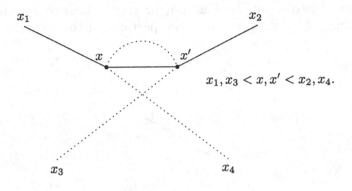

$$x_1, x_3 < x, x' < x_2, x_4.$$

Fig. 1.7

This diagram describes a process which is essentially different from the
first one. The new diagram cannot be reduced to that in Fig. 1.6 by
moving the interaction points: the two pictures are topologically different. Our process can go now via two independent routes. Therefore, the
expressions corresponding to Figs. 1.6 and 1.7 should be *added* as the
amplitudes of independent quantum processes to give the full transition
amplitude $1 + 2 \to 1 + 2$ in the second order in λ.

The particles attached to points x_3 and x_4 are identical. The question
is whether our scattering amplitude is aware of it. As we have already
explained, due to relativistic invariance we can freely play with the time
coordinates, looking at the diagrams with different ordering of times in-
volved. Let us choose

$$x_1,\ x_2\ <\ x_3,\ x_4\,,$$

which is equivalent to looking at the graphs Figs. 1.6 and 1.7 'from the top'
(changing the time-arrow from the accepted horizontal to vertical). By
doing so we come to the transition process $1+1 \to 2+2$ with two identical
particles in the final state. The total amplitude we have obtained by
summing Figs. 1.6 and 1.7 is then automatically *symmetric* with respect
to transmutation of particle coordinates: under the replacement $x_3 \leftrightarrow x_4$
the diagrams Figs. 1.6 and 1.7 interchange, and their sum remains the
same. (This is a hint for the future relation between particle spin and
statistics.)

Having designed the fundamental interaction we can now multiply the particles in arbitrary numbers:

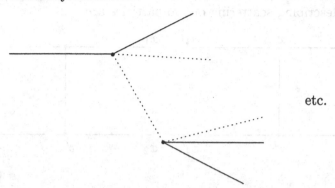

etc.

We have considered a theory with two species of particles merely for convenience, to make it easier to see how the diagrams transformed under different time orderings. We can construct the relativistic interaction having only one sort of particle at our disposal. Again, we may take four-particle interaction as the primitive process, which contains scattering and decay configurations:

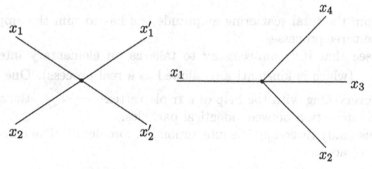

Is a three-particle interaction

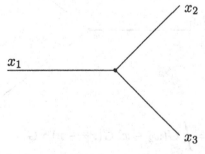

also possible? As a real process, the decay of a particle into two is forbidden by energy–momentum conservation. For finite time intervals, however, its amplitude is different from zero and we can use it to describe scattering of real particles which interact by exchanging a *virtual* particle.

In the lowest order (two interaction vertices) the topologically different diagrams describing scattering of two particles are

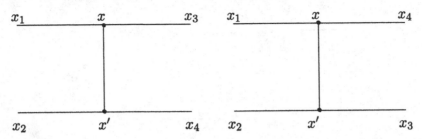

and

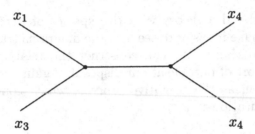

To obtain the total scattering amplitude one has to sum the amplitudes of these three processes.

We see that it is unnecessary to take as an elementary interaction ✕ (which is kinematically allowed as a real process). One can describe everything with the help of a triple vertex ─< , the simplest kind of interaction between identical particles.

Let us study three-particle interaction in more detail. The corresponding amplitude is

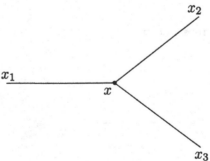

$$G(x_2, x_3; x_1) = \int G(x_2 - x)\, G(x_3 - x)\, \gamma\, G(x - x_1)\, d^4x \,. \qquad (1.99)$$

Inserting the Fourier representation (1.96) for the Green functions, we get

$$G(x_2, x_3; x_1) = \gamma \int \frac{d^4p_1 d^4p_2 d^4p_3}{[(2\pi)^4 i]^3} \frac{1}{(m^2 - p_1^2)(m^2 - p_2^2)(m^2 - p_3^2)}$$

$$\times \int d^4x e^{-ip_2(x_2-x)-ip_3(x_3-x)-ip_1(x-x_1)}$$

$$= \gamma \int \frac{d^4p_1 d^4p_2 d^4p_3 (2\pi)^4 \delta(p_2+p_3-p_1)}{[(2\pi)^4 i]^3 (m^2-p_1^2)(m^2-p_2^2)(m^2-p_3^2)} e^{-ip_2x_2-ip_3x_3+ip_1x_1}$$

$$\equiv \int \frac{d^4p_1 d^4p_2 d^4p_3}{[(2\pi)^4 i]^3} (2\pi)^4 \delta(p_2+p_3-p_1) G(p_1,p_2,p_3) e^{-ip_2x_2-ip_3x_3+ip_1x_1},$$

where $\delta(p_2 + p_3 - p_1)$ reflects the energy–momentum conservation. The Green function in momentum space is extremely simple:

$$G(p_1,p_2,p_3) = \gamma\, G(p_1)\, G(p_2)\, G(p_3). \qquad (1.100)$$

The corresponding diagram in momentum space is

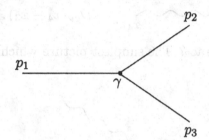

The expression (1.100) describes virtual particles, since the on-mass-shell conditions $p_i^2 = m^2$ cannot be satisfied by all three momenta simultaneously. As a result, the corresponding amplitude $G(x_2, x_3; x_1)$ in coordinate space vanishes exponentially in the limit $t_1 \to -\infty$ and $t_2, t_3 \to +\infty$ which would correspond to the real decay of a particle into two.

To prove this, let us integrate over p_3 in (1.99). We obtain

$$G(x_2, x_3; x_1)$$
$$= \gamma \int \frac{d^4p_2 d^4p_1}{[(2\pi)^4 i]^3} (2\pi)^4 \frac{e^{ip_2(x_3-x_2)+ip_1(x_1-x_3)}}{(m^2-p_1^2)(m^2-p_2^2)[m^2-(p_1-p_2)^2]}. \qquad (1.101)$$

Note that in the region of integration in (1.101) there is no point where all three denominators vanish. As a result the integral at $t_1 \to -\infty$ and $t_2, t_3 \to \infty$ cannot be written in the form of a momentum space integral of the product of the initial and final real particle wave functions and a finite momentum space decay amplitude.§ This means that at large initial and final times all three particles cannot be real simultaneously and the real decay cannot occur.

The possibility of a real decay of a particle into two is governed by energy–momentum conservation. For identical particles, such a process is

---

§ Compare with the discussion of the S-matrix and especially equation (1.131) below.

forbidden, though a sufficiently heavy particle 1 can decay into two real light particles 2 and 3 provided $m_1 > m_2 + m_3$.

1.8 Interaction of spinless particles with the electromagnetic field

Now consider the interaction of a charged particle with a photon. We know the free Green functions of electrons and photons:

$$G(x_2 - x_1)$$

$$D_{\mu\nu}(x_4 - x_3).$$

How might they interact? The simplest picture which comes to mind is

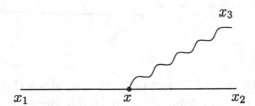

As we already know, it is not a *real* process since the corresponding amplitude vanishes in the limit $x_1 \to -\infty$, $x_2, x_3 \to +\infty$. Still, it can be taken as the basic building block for constructing the interaction between charged particles and photons.

The photon is a vector particle and its interaction may depend on the photon polarization. Therefore, the interaction amplitude Γ_ν should bear the vector index ν. The four-vector amplitude for the emission of a photon takes the form

$$G_\mu(x_3, x_2; x_1) = \int D_{\mu\nu}(x_3 - x)\, G(x_2 - x)\, \Gamma_\nu(x)\, G(x - x_1) d^4x. \quad (1.102)$$

As before, Γ_ν should not depend on the position x_μ. On the other hand, it is a four-vector, and the only such vector we can invent besides x_μ is $\partial/\partial x_\mu$. Hence,

$$G(x_2 - x)\Gamma_\nu(x)G(x - x_1)$$
$$= a\, G(x_2 - x)\frac{\partial G(x - x_1)}{\partial x_\nu} + b\frac{\partial G(x_2 - x)}{\partial x_\nu}G(x - x_1), \quad (1.103)$$

where a and b are arbitrary constants. (We do not need to differentiate the photon Green function $D_{\mu\nu}(x_3 - x)$ because this derivative can always be traded for differentiation of the functions G using integration by parts in (1.102).)

We have to impose an additional condition on the interaction to exclude the two unphysical polarizations which are present in $D_{\mu\nu} \propto g_{\mu\nu}$. The electromagnetic field has to satisfy the Lorentz condition at the point x_3 (at least for $x_3 \neq x_1, x_2$, see below), and therefore we should have

$$\frac{\partial G_\mu(x_3, x_2; x_1)}{\partial x_{3\mu}} = 0. \tag{1.104}$$

Since the amplitude G_μ depends on x_3 only via $D_{\mu\nu}$, it suffices to differentiate $D_{\mu\nu}$ in (1.104), which gives

$$\frac{\partial D_{\mu\nu}(x_3 - x)}{\partial x_{3\mu}} = i \int \frac{d^4 k}{(2\pi)^4 i} \frac{e^{-ik(x_3 - x)}}{k^2} k_\mu \sum_{\lambda=0}^{3} e_\mu^\lambda e_\nu^{\lambda *}.$$

The condition $k_\mu e_\mu^\lambda = 0$ is valid only for three vectors e^λ ($\lambda = 1, 2, 3$), while $k_\mu e_\mu^{(0)} \neq 0$. To satisfy the Lorentz condition at x_3, the contribution of the scalar polarization $e^{(0)}$ should disappear from (1.104). Effectively, this is a condition on the interaction vertex Γ_ν imposed by current conservation.

Let us calculate the divergence $\partial G_\mu / \partial x_{3\mu}$ starting from the definition of the amplitude given in (1.102) ($D_{\mu\nu} \equiv g_{\mu\nu} D(x)$).

$$\frac{\partial G_\mu(x_3, x_2; x_1)}{\partial x_{3\mu}} = g_{\mu\nu} \int \frac{\partial D(x_3 - x)}{\partial x_{3\mu}} G(x_2 - x) \Gamma_\nu G(x - x_1) \, d^4 x$$

$$= \int D(x_3 - x) \frac{\partial}{\partial x_\mu} \Big(G(x_2 - x) \Gamma_\mu G(x - x_1) \Big) d^4 x. \tag{1.105}$$

Here we used $\partial D(x_3 - x)/\partial x_{3\mu} = -\partial D(x_3 - x)/\partial x_\mu$ and integrated by parts.

The expression (1.105) equals zero for arbitrary x_3 provided that

$$\frac{\partial}{\partial x_\mu} \Big(G(x_2 - x) \, \Gamma_\mu \, G(x - x_1) \Big) = 0. \tag{1.106}$$

Thus, we have a condition on the interaction. To specify it, we insert (1.103) into (1.106) to get

$$\frac{\partial}{\partial x_\mu} \Big(G(x_2 - x) \Gamma_\mu(x) G(x - x_1) \Big)$$

$$= a \frac{\partial G(x_2 - x)}{\partial x_\mu} \frac{\partial G(x - x_1)}{\partial x_\mu} + b \frac{\partial G(x_2 - x)}{\partial x_\mu} \frac{\partial G(x - x_1)}{\partial x_\mu}$$

$$+ a \, G(x_2 - x) \frac{\partial^2 G(x - x_1)}{\partial x_\mu^2} + b \frac{\partial^2 G(x_2 - x)}{\partial x_\mu^2} G(x - x_1).$$

Because $\partial^2 G(x)/\partial x_\mu^2 = -m^2 G - i\delta(x)$, the terms containing masses will disappear together with the first two terms on the right-hand side, if we put $a = -b$. The remaining piece,

$$\frac{\partial}{\partial x_\mu}(G\Gamma_\mu G) = a\left[G(x_2 - x_1)(-i)\delta(x_1 - x) - G(x_2 - x_1)(-i)\delta(x - x_2)\right],$$

differs from zero only when either $x = x_1$ or $x = x_2$ and corresponds to photon emission at the very moment of particle creation (absorption). Strictly speaking, one should consider these situations separately, adding to the main process pictures of the type:

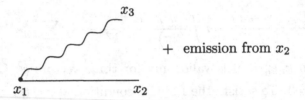

$+$ emission from x_2

in order to preserve the current conservation exactly. This additional piece, however, never enters physical processes. The charged particles one studies in real experiments are prepared long in advance so that the photons which might have been created then do not affect the measurement (do not hit detectors).

So, our best choice for the photon emission vertex is

$$\Gamma_\mu(x) = \gamma \frac{\overleftrightarrow{\partial}}{\partial x_\mu} \tag{1.107}$$

and, respectively,

$$G_\mu(x_3, x_2; x_1) = \gamma \int D(x_3 - x)\left(G(x_2 - x)\frac{\overleftrightarrow{\partial}}{\partial x_\mu}G(x - x_1)\right)d^4x. \tag{1.108}$$

We shall associate this amplitude with the diagram given in Fig. 1.8.

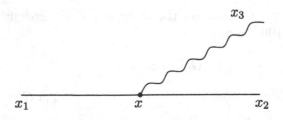

Fig. 1.8

The theory we have developed so far is not quite satisfactory. Consider the region of integration in (1.108) where $t_1, t_2 < t_3$. This corresponds to the graph given in Fig. 1.9,

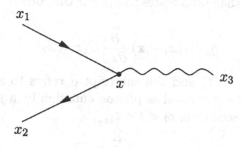

Fig. 1.9

which describes two charged particles merging into a photon. What happened to the electric charge which, as we know from classical physics, should be conserved? The picture makes no sense.

There is also another contradiction, a formal one. Before, we were forced by relativistic invariance to suppose that the Green function G propagates the same particle forward and backward in time. However, we know that the amplitude of a process involving identical spin-zero particles should be symmetric under their permutation (Bose statistics). Meanwhile, the amplitude we obtained in (1.108) is apparently antisymmetric under transposition $x_1 \leftrightarrow x_2$ (due to the antisymmetry of the operator $\overleftrightarrow{\partial}/\partial x_\mu$).

The way out of these contradictions is to make the *hypothesis* that for charged particles there is always degeneracy: for every charged particle there exists an *antiparticle* with the same mass. Since the Green functions of particles with equal masses coincide, this assumption enables us to give a different interpretation of the graph in Fig. 1.9, according to which an antiparticle rather than a particle is propagating from x_2. In other words, $G(x)$ describes the propagation of a particle if $t > 0$ and that of a different particle (antiparticle) if $t < 0$.

Now, by ascribing to the antiparticle an electric charge equal and opposite to that of the particle, we can rescue charge conservation. The diagram in Fig. 1.9 now describes a legitimate process of annihilation of two particles with opposite charges into a photon.

What about the second problem? It seems to have disappeared, since the two charged particles in Fig. 1.9 are no longer identical, so that we do not need to bother about permuting their coordinates. Nevertheless, the antisymmetry of the amplitude gives us important information about the interaction constant γ we introduced in the interaction vertex (1.107): the coupling constant changes sign when we replace a particle by its antiparticle. This hints at the future identification of γ with the electric charge of the particle emitting a photon.

Indeed, let us consider an antiparticle which propagates from x_2, emits a photon at x and then propagates to x_1. For the corresponding amplitude we would write

$$\gamma_a \cdot G(x_1 - x) \frac{\overset{\leftrightarrow}{\partial}}{\partial x_\mu} G(x - x_2) \, ,$$

where we used $G_a = G$ and the subscript a refers to antiparticle. The same process can be described as photon emission by a *particle* by taking the reversed time sequence, $t_2 < t < t_1$,

$$\gamma \cdot G(x_2 - x) \frac{\overset{\leftrightarrow}{\partial}}{\partial x_\mu} G(x - x_1) \, .$$

The latter expression differs from the previous one only in the order in which the differentiation of the functions G is carried out. Therefore,

$$\gamma_a = -\gamma \, .$$

The operator $\overset{\leftrightarrow}{\partial}_\mu$ is rather inconvenient to use for the photon emission vertex because we always have to keep track of the time ordering (which G stands on the right and which on the left from $\overset{\leftrightarrow}{\partial}_\mu$) and whether the amplitude is written for a particle or its antiparticle (the sign of γ). The annihilation diagram in Fig. 1.9 is the most confusing: which order for the two Green functions in $G \overset{\leftrightarrow}{\partial}_\mu G$ to prefer and, correspondingly, which factor (γ or $\gamma_a = -\gamma$) to supply it with?

It is useful to introduce arrows in the graphs and change the previous convention $\overset{\leftrightarrow}{\partial} = \overset{\rightarrow}{\partial} - \overset{\leftarrow}{\partial}$ to the new prescription shown in Fig. 1.10.

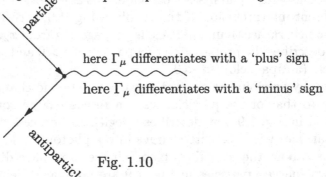

here Γ_μ differentiates with a 'plus' sign

here Γ_μ differentiates with a 'minus' sign

Fig. 1.10

Supplying the charged particle line with an arrow provides a convenient way to distinguish between particles and antiparticles. This enables us to see immediately if the differentiation is carried out with a plus or a minus sign irrespective of how the graph is oriented (that is, independently of the time sequence of events). Within this new convention, photon emission by a particle and its antiparticle is described by expression (1.104) with the same parameter γ.

Existence of antiparticles, which has been confirmed by numerous experiments, is not a law of nature in itself. We have predicted antiparticles following the conservation of charge (a certain form of the electromagnetic interaction) and relativistic invariance. This prediction fits into the general pattern: Conservation → Symmetry → Degeneracy. In general, existence of antiparticles is always a consequence of conservation of some kind of charge (not necessarily electric). For example, existence of anti-neutrons is connected with baryon charge conservation, that of anti-K mesons with conservation of hypercharge, etc.

1.9 Examples of the simplest electromagnetic processes

For the time being we shall treat the simplest processes with the minimal number of interactions between photons and charged particles. But before that let us briefly discuss the rôle of *higher orders*.

In addition to the process ⌇⌇ we have considered above, photon emission could also occur as a result of more complicated processes, for example,

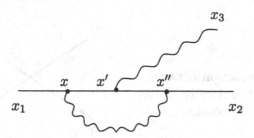

Can we restrict ourselves to the simplest processes? Yes, if the probability of more complicated ones is relatively small. To this end, let us compare the amplitudes corresponding to the two graphs above.

The first amplitude is described by (1.108). For the second one we have

$$A'_\mu(x_3, x_2; x_1) = \int d^4x d^4x' d^4x'' \, D(x'' - x) D(x_3 - x')$$

$$\times \, G(x_2 - x'')\Gamma_\nu(x'')G(x'' - x')\Gamma_\mu(x')G(x' - x)\Gamma_\nu(x)G(x - x_1) \, .$$

Compared to (1.108), here there are two more Green functions, two more vertices Γ_μ, one more D and two more integrations. The dimension of this 'extra' part is

$$G \, G \, D \, \Gamma \, \Gamma \, d^4x \, d^4x \sim \frac{1}{x^2} \frac{1}{x^2} \frac{1}{x^2} \frac{\gamma}{x} \frac{\gamma}{x} x^8 \sim \gamma^2 \, .$$

Since the dimensions of both amplitudes should be the same, γ must be dimensionless. Hence, it suffices to have $\gamma^2 \ll 1$ to allow us to ignore more complicated processes and consider the simplest diagrams only.

Let us consider the simplest processes with real charged particles and their diagrammatic description.

1.9.1 Scattering of charged particles

As we have already discussed, 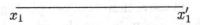 does not correspond to any real physical process (a free electron cannot emit a real photon). What can happen with two charged particles?

(a) The particles do not interact

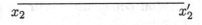

(b) Contact interaction between particles without emission or absorption of photons:

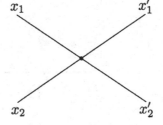

Whether such an interaction exists is an experimental question. It does exist for some particles, and does not for others. (For example, contact interaction exists for pions.)

Existing or not, we shall ignore it anyway, because it has nothing to do with the emission or absorption of photons we are interested in. We will consider only electromagnetic interactions, i.e. we will assume that non-electromagnetic contact interaction is absent. (In fact, the field theory we are about to construct will be a simplified quantum electrodynamics of electrons and muons without spin.)

(c) Scattering of charged particles via photon exchange. The simplest diagrams with one photon exchange are shown in Fig 1.11.

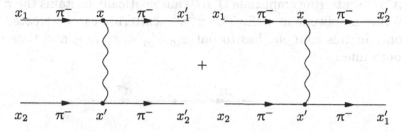

Fig. 1.11

Both of these processes are possible, since after the emission of a photon at x the particle from the point x_1 can propagate to x_1' as well as to x_2'. The amplitude can be written as

$$G(x_2', x_1'; x_2, x_1) = \int d^4x\, d^4x'\, G(x_1' - x)\Gamma_\mu G(x' - x)$$
$$\times\, G(x_2' - x')\Gamma_\nu G(x' - x_2) + \{x_1' \leftrightarrow x_2'\},$$
$$(1.109)$$

where $\{x_1' \leftrightarrow x_2'\}$ denotes an expression identical to the previous one but with x_1' and x_2' transposed.

From the first diagram in Fig. 1.11 follows the existence of the graph

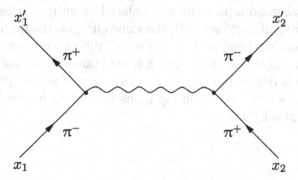

from the second diagram that of

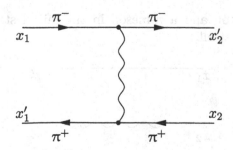

These graphs correspond to the process of $\pi^+\pi^-$ scattering. Hence, the $\pi^-\pi^-$ scattering amplitude (1.109) automatically contains the $\pi^+\pi^-$ scattering amplitude as well. The $\pi^+\pi^+$ scattering can be treated similarly, only in this case one has to put $x'_{10}, x'_{20} < x_{10}, x_{20}$ and thus to reverse both lines:

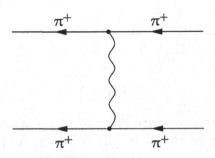

Thus, once we have written the amplitude of $\pi^-\pi^-$ scattering, we automatically obtain the $\pi^-\pi^+$ and $\pi^-\pi^+$ processes. This is a powerful consequence of degeneracy, which itself resulted from relativity. The amplitudes of these new processes follow immediately from (1.109). We only have to choose the initial and final times properly, that is, to redirect the lines of the original diagrams in Fig. 1.11.

The $\pi^+\pi^+ \to \pi^+\pi^+$ amplitude is identical to the $\pi^-\pi^- \to \pi^-\pi^-$ amplitude, because the amplitude contains two factors Γ_μ and each of them changes sign when both particles are replaced by antiparticles. Unlike the case of scattering of particles with the same charges, the amplitude of the particle–antiparticle process $\pi^+\pi^- \to \pi^+\pi^-$ differs significantly from the two previous amplitudes, even though it is obtained from the same initial formula. This is due to the presence in this case of a new, physically different, process where two incoming pions annihilate in the intermediate state into a virtual photon.

1.9.2 The Compton effect (photon-π-meson scattering)

Let us take a photon and a π-meson in the initial state. Again, they might not interact at all:

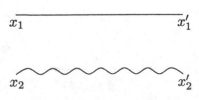

If they do interact, the meson can absorb the photon at a point x and emit it at x':

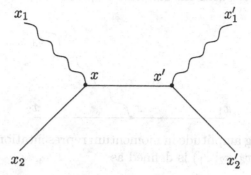

Alternatively, the meson can emit a photon at a point x and absorb the initial photon at x':

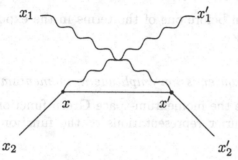

Is a contact interaction possible as well?

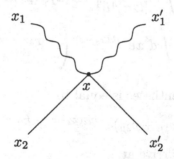

As we have discussed above (see Section 1.9.1), there is no a priori answer to this question. In general, its existence is an experimental problem. In electrodynamics, however, the situation is special. Here the interaction should satisfy conditions imposed by current conservation which forbids contributions of scalar and real longitudinal photons to any physical observables. It turns out that, indeed, in the case of scalar charged particles current conservation requires the presence of such contact interaction. As we shall see later, the strength of this contact interaction is proportional to γ^2.

1.10 Diagrams and amplitudes in momentum representation

We begin, as before, with the simplest diagram for photon–meson inter-action.

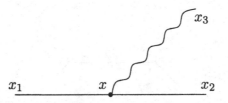

The corresponding amplitude in momentum representation, i.e. the Fourier transform of $G_\mu(x_3, x_2; x_1)$ is defined as

$$G_\mu(x_3, x_2; x_1) = \int \frac{d^4 p_1 d^4 p_2 d^4 k}{[(2\pi)^4 i]^3} e^{-ip_1 x_1 + ip_2 x_2 + ik x_3} G_\mu(p_1, p_2, k), \quad (1.110)$$

where the minus sign before one of the terms in the exponent is chosen merely for convenience.

1.10.1 Photon emission amplitude in momentum space

Let us now calculate the momentum-space Green function $G_\mu(p_1, p_2, k)$. Substituting the Fourier representations of the functions G and D in (1.108) we obtain

$$G_\mu(x_3, x_2; x_1) = \gamma \int \frac{d^4 p_1 d^4 p_2 d^4 k}{[(2\pi)^4 i]^3} D(k) G(p_1) G(p_2)$$

$$\times \int d^4 x e^{-ik(x_3 - x)} \left(e^{-ip_2(x_2 - x)} \frac{\overleftrightarrow{\partial}}{\partial x_\mu} e^{-ip_1(x - x_1)} \right).$$
$$(1.111)$$

The expression in parentheses is equal to

$$-i(p_{1\mu} + p_{2\mu}) e^{-ip_2(x_2 - x)} e^{-ip_1(x - x_1)}.$$

Integrating over x we arrive at

$$G_\mu(x_3, x_2; x_1) = -\gamma \int \frac{d^4 p_1 d^4 p_2 d^4 k}{[(2\pi)^4 i]^3} D(k) G(p_1) G(p_2)$$

$$\times i(p_{1\mu} + p_{2\mu})(2\pi)^4 \delta(p_1 - p_2 - k) e^{ip_1 x_1 - ip_2 x_2 - ik x_3}.$$
$$(1.112)$$

Finally, comparing (1.110) and (1.112) we derive

$$G_\mu(p_1, p_2, k) = -(2\pi)^4 i\delta(p_1 - p_2 - k)(p_{1\mu} + p_{2\mu})\gamma\, G(p_1) G(p_2) D(k).$$
$$(1.113)$$

The corresponding diagram is

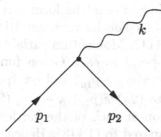

where the lines correspond to the Green functions $G(p_1)$, $G(p_2)$, $D(k)$, the vertex Γ_μ corresponds to $-i\gamma(p_{1\mu}+p_{2\mu})$, and the factor $(2\pi)^4\delta(p_1-p_2-k)$ is due to the energy–momentum conservation. Thus we see that the rules which connect graphs and amplitudes in momentum space are simpler than in coordinate space.

This particular diagram describes the 'decay' of a meson with momentum p_1 into a photon with momentum k and a meson with momentum $p_2 = p_1 - k$. In this process the four-momenta are *off* the mass shell: $p_0^2 \neq \mathbf{p}^2 + m^2$ and/or $k_0^2 \neq \mathbf{k}^2$. Such processes cannot occur for real particles and are called virtual processes.

1.10.2 Meson–meson scattering via photon exchange

Consider now meson–meson scattering

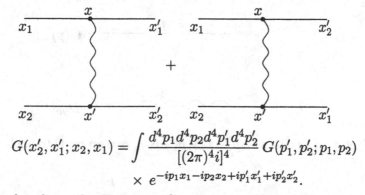

$$G(x_2', x_1'; x_2, x_1) = \int \frac{d^4p_1 d^4p_2 d^4p_1' d^4p_2'}{[(2\pi)^4 i]^4}\, G(p_1', p_2'; p_1, p_2)$$
$$\times\, e^{-ip_1x_1 - ip_2x_2 + ip_1'x_1' + ip_2'x_2'}.$$

We can also draw the Feynman diagrams in momentum space. For example, for the first diagram we have

$$p_1 \quad \gamma(p_{1\mu} + p_{1\mu}') \quad p_1'$$

$$k$$

$$p_2 \quad \gamma(p_{2\mu} + p_{2\mu}') \quad p_2'$$

In order to obtain the momentum-space Green function we proceed as above. We write $G(x_2', x_1'; x_2, x_1)$ in the form (1.109), and substitute all free Green functions by their Fourier representations. The resulting formula is of the type of (1.111). Momentum variables originating from the Fourier transforms correspond to each Green function. In the vertices the differentiations give $-i\gamma(p_{1\mu} + p_{1\mu}')$ and $-i\gamma(p_{2\mu} + p_{2\mu}')$. The integrations over x and x' lead to $(2\pi)^4\delta(p_1 + k - p_1')$, $(2\pi)^4\delta(p_2 - k - p_2')$, i.e. to momentum conservation at each of the vertices. There is one extra integration over k as compared to (1.113). Hence,

$$G(p_1', p_2'; p_1, p_2) = \int \frac{d^4k}{(2\pi)^4 i} D(k)\, G(p_1)\, G(p_2)\, G(p_1')\, G(p_2')$$
$$\times \gamma^2 (p_1 + p_1')_\mu (p_2 + p_2')_\mu (2\pi)^4 i\delta(p_1 + k - p_1')\, (2\pi)^4 i\delta(p_2 - k - p_2').$$

$$(1.114)$$

The expression (1.114) contains an integration over the momentum k of the intermediate photon.

1.10.3 Feynman rules

Similarly, we can formulate the rules for constructing the amplitudes corresponding to arbitrary diagrams in momentum space.

(1) A multiplicative factor (a Green function) corresponds to each line:

$$\underline{\qquad\qquad p \qquad\qquad}\qquad G(p)$$

$$\sim\!\!\sim\!\!\sim\!\!\sim^{\;k}\!\!\sim\!\!\sim\qquad D_{\mu\nu}(k)$$

(2) A factor $-(2\pi)^4 i\delta(p_1 - p_2 - k)\gamma(p_1 + p_2)_\mu$ corresponds to each vertex:

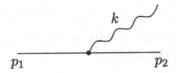

(3) One has to integrate over the momenta of the intermediate particles (with the weight $d^4k/(2\pi)^4 i$), i.e. over the momenta which correspond to the internal lines.

Let us return to the expression (1.114) for the two-particle scattering amplitude. The δ-function takes care of the momentum integration on the right-hand side. We also have to take into account the second diagram which in momentum space has the form

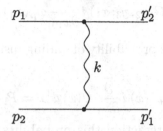

As a result the total amplitude can be written as

$$G(p'_2, p'_1; p_2, p_1) = (2\pi)^4 i\delta(p_1 + p_2 - p'_1 - p'_2)G(p_1)G(p_2)G(p'_1)G(p'_2)$$
$$\times \gamma^2 \left[\frac{(p_1+p'_1)_\mu(p_2+p'_2)_\mu}{(p_2 - p'_2)^2} + \frac{(p_1+p'_2)_\mu(p_2+p'_1)_\mu}{(p_2 - p'_1)^2} \right], \quad (1.115)$$

where the δ-function reflects the energy–momentum conservation in the scattering process.

1.11 Amplitudes of physical processes

Let us assume that at time t_1 a particle is described by the wave function[¶]

$$\varphi_{\mathbf{p}}(x) = \frac{e^{-ipx}}{\sqrt{2E}}. \quad (1.116)$$

Then, at another moment in time it will be described by the function

$$\Psi(y_1) = \int G(y_1 - x_1)\, i\frac{\overleftrightarrow{\partial}}{\partial x_{10}}\, \varphi_{\mathbf{p}}(x_1)\, d^3 x_1, \quad (1.117)$$

where

$$G(y_1 - x_1) = \frac{}{x_1 \hspace{4cm} y_1}.$$

A virtual process can occur in which this particle would decay into two:

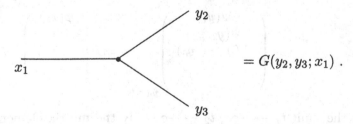

$$= G(y_2, y_3; x_1).$$

[¶] For simplicity we shall consider the case when there exists only one species of particles.

In this case the wave function of the system will be

$$\Psi(y_2, y_3) = \int G(y_2, y_3; x_1)\, i\frac{\overleftrightarrow{\partial}}{\partial x_{10}}\, \varphi_{\mathbf{p}}(x_1)\, d^3 x_1\,. \qquad (1.118)$$

As already explained, the probability of finding just one particle anywhere in space is

$$\int \varphi^*(x)\, i\frac{\overleftrightarrow{\partial}}{\partial x_0}\, \Psi(x)\, d^3 x = P_1\,.$$

In the presence of an interaction this probability is less than one since there are certain probabilities P_2, P_3 etc. to find two or more particles, for example

$$P_2 = \int \varphi^*_{\mathbf{p}_1}(x_1)\varphi^*_{\mathbf{p}_2}(x_2)\, i\frac{\overleftrightarrow{\partial}}{\partial x_{10}}\, i\frac{\overleftrightarrow{\partial}}{\partial x_{20}}\, \Psi(x_1, x_2)\, d^3 x_1\, d^3 x_2\,. \qquad (1.119)$$

According to the orthogonality condition (1.56), the probability amplitude for a transition of a particle with momentum $\mathbf{p}_1$ into a state with momentum $\mathbf{p}_2$ is

$$\int \varphi^*_{\mathbf{p}_2}(y)i\frac{\overleftrightarrow{\partial}}{\partial y_0}\Psi(y)d^3 y = \int \varphi^*_{\mathbf{p}_2}(y)i\frac{\overleftrightarrow{\partial}}{\partial y_0}G(y-x)i\frac{\overleftrightarrow{\partial}}{\partial x_0}\varphi_{\mathbf{p}_1}(x)d^3 x d^3 y. \qquad (1.120)$$

The probability amplitude for the decay of this particle into two particles with momenta $\mathbf{p}_2$ and $\mathbf{p}_3$ is as follows:

$$\langle \mathbf{p}_2, \mathbf{p}_3 | \mathbf{p}_1 \rangle = \int \varphi^*_{\mathbf{p}_2}(y_2)\varphi^*_{\mathbf{p}_3}(y_3)\, i\frac{\overleftrightarrow{\partial}}{\partial y_{20}}i\frac{\overleftrightarrow{\partial}}{\partial y_{30}}G(y_3, y_2; x_1)$$

$$\times\, i\frac{\overleftrightarrow{\partial}}{\partial x_0}\varphi_{\mathbf{p}_1}(x_1)\, d^3 x_1 d^3 y_2 d^3 y_3\,. \qquad (1.121)$$

Hence, we can define a matrix U which transforms an initial one-particle state $\varphi(x_1)$ at time t_1 into all possible states at t_2, i.e.

$$\begin{pmatrix} \Psi(y_1) \\ \Psi(y_2, y_3) \\ \Psi(y_4, y_5, y_6) \\ \cdot \\ \cdot \end{pmatrix}_{y_{i0}=t_2} = U \begin{pmatrix} \varphi(x_1) \\ 0 \\ 0 \\ \cdot \\ \cdot \end{pmatrix}_{t_1}. \qquad (1.122)$$

In the limit $t_1 \to -\infty$, $t_2 \to +\infty$ only the matrix element U_{11} does not vanish since the decay of a physical particle is forbidden by conservation laws.

1.11.1 The unitarity condition

If at the initial moment there were two particles, $\varphi(x_1, x_2)$, then, in analogy with (1.122),

$$
\begin{pmatrix} \Psi(y_1) \\ \Psi(y_2, y_3) \\ \Psi(y_4, y_5, y_6) \\ \cdot \\ \cdot \end{pmatrix}_{y_{i0}=t_2} = U \begin{pmatrix} 0 \\ \varphi(x_1, x_2) \\ 0 \\ \cdot \\ \cdot \end{pmatrix}_{x_{i0}=t_1} . \tag{1.123}
$$

In this case there are many possible processes in the limit $t_1 \to -\infty$, $t_2 \to +\infty$, because two particles can create new particles and scatter as well. The only requirement is that the total probability of all processes has to be unity. Hence it follows that the operator U must be unitary, i.e.

$$
U^+(t_2, t_1)\, U(t_2, t_1) = I . \tag{1.124}
$$

Since we consider the case when the coupling constant is small, $\gamma \ll 1$, the matrix U must be very close to unity, i.e. it is natural to write it in the form

$$
U = I + iV . \tag{1.125}
$$

The relation (1.124) gives $-iV^+ + iV = \mathcal{O}(\gamma^2) \ll |V|$, or

$$
V^+ \simeq V . \tag{1.126}
$$

This means that the term in U which is induced by the interaction has to be imaginary. (Recall that we had $-iV$ in our Feynman rules for non-relativistic quantum mechanics. Note also that there is a factor i in the photon emission vertex $\Gamma_\mu = -i\gamma(p_1 + p_2)_\mu$. We can expect, therefore, that γ will turn out to be real.)

1.11.2 S-matrix

We define the scattering matrix S, or simply S-matrix, as the limit $t_1 \to -\infty$, $t_2 \to +\infty$,

$$
S = \lim_{t_1 \to -\infty, t_2 \to \infty} U(t_2, t_1) . \tag{1.127}
$$

The probability amplitudes for different processes can be calculated similarly to (1.120) and (1.121). For example, for a scattering process (two particles → two particles) we have

$$
S(\mathbf{p}_3, \mathbf{p}_4 | \mathbf{p}_1, \mathbf{p}_2) = \int \varphi_{\mathbf{p}_3}^*(y_1) \varphi_{\mathbf{p}_4}^*(y_2) i\frac{\overleftrightarrow{\partial}}{\partial y_{20}} i\frac{\overleftrightarrow{\partial}}{\partial y_{10}} G(y_2, y_1; x_2, x_1)
$$

$$
\times\; i\frac{\overleftrightarrow{\partial}}{\partial x_{10}} i\frac{\overleftrightarrow{\partial}}{\partial x_{20}} \varphi_{\mathbf{p}_1}(x_1)\varphi_{\mathbf{p}_2}(x_2)\, d^3 x_1 d^3 x_2 d^3 y_1 d^3 y_2 .
$$

There is a more transparent method of calculation. The Green function for a process with a given number of particles before and after the interaction can be represented by the diagram

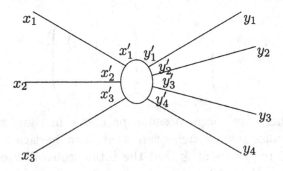

An arbitrary process can be drawn like this, where the bubble stands for all possible intermediate states.

The only difference between this diagram and the diagrams for the Green functions we have discussed above is that now

$$x_{i0} \to -\infty, \quad \text{i.e.} \quad x'_{i0} - x_{i0} > 0,$$

$$y_{i0} \to +\infty, \quad \text{i.e.} \quad y_{i0} - y'_{i0} > 0. \tag{1.128}$$

These conditions allow us to simplify the free Green functions for the external lines. Indeed, (1.128) determines unambiguously how to close the contours in the integrals. For example

$$G(y_1 - y'_1)\big|_{y_{10}-y'_{10}>0} = \int \frac{d^4 p_1}{(2\pi)^4 i} \frac{e^{-ip_1(y_1-y'_1)}}{m^2 - p_1^2 - i\varepsilon} = \int \frac{d^3 p_1}{(2\pi)^3} \frac{e^{-ip_1(y_1-y'_1)}}{2E_1},$$

$$\tag{1.129}$$

where $E_1 = \sqrt{\mathbf{p}_1^2 + m^2}$, i.e. the particle is real. On the other hand, using (1.95) the expression (1.129) can be written as

$$G(y_1 - y'_1) = \int \frac{d^3 p_1}{(2\pi)^3} \varphi^*_{\mathbf{p}_1}(y'_1)\, \varphi_{\mathbf{p}_1}(y_1). \tag{1.130}$$

Hence, an arbitrary Green function $x_{i0} \to -\infty$, $y_{i0} \to +\infty$ can be represented in the form

$$G(y_1, y_2, \ldots y_n; x_1, x_2, \ldots x_m)$$
$$= \int \frac{d^3 p_1 \ldots d^3 p_n}{(2\pi)^{3n}} \varphi_{\mathbf{p}_1}(y_1) \ldots \varphi_{\mathbf{p}_n}(y_n) \int \frac{d^3 k_1 \ldots d^3 k_m}{(2\pi)^{3m}} \varphi^*_{\mathbf{k}_1}(x_1) \ldots \varphi^*_{\mathbf{k}_m}(x_m)$$
$$\times \int d^4 y'_1 \ldots d^4 y'_n \varphi^*_{\mathbf{p}_1}(y'_1) \ldots \varphi^*_{\mathbf{p}_n}(y'_n) \int d^4 x'_1 \ldots d^4 x'_m \varphi_{\mathbf{k}_1}(x'_1) \ldots \varphi_{\mathbf{k}_m}(x'_m)$$
$$\times S(y'_1, \ldots y'_n; x'_1, \ldots x'_m).$$

Thus, the Green function is a superposition of plane waves,

$$
G(y_1, y_2, \ldots y_n; x_1, x_2, \ldots x_m)
$$
$$
= \prod_{i=1}^{n} \int \frac{d^3 p_i}{(2\pi)^3} \varphi_{\mathbf{p}_i}(y_i) \prod_{j=1}^{m} \frac{d^3 k_j}{(2\pi)^3} \varphi^*_{\mathbf{k}_j}(x_j) \cdot S(p_1, \ldots p_n; k_1, \ldots k_m).
$$

The weight of this superposition is just the transition amplitude between the initial and final states with given momenta

$$
S(p_1, \ldots p_n; k_1, \ldots k_m)
$$
$$
= \prod_{i=1}^{n} \int d^4 y'_i \, \varphi^*_{\mathbf{p}_i}(y'_i) \prod_{j=1}^{m} \int d^4 x'_j \varphi_{\mathbf{k}_j}(x'_j) \, S(y'_1, \ldots y'_n; x'_1, \ldots x'_m). \quad (1.131)
$$

From (1.131) it becomes clear that the transition amplitude for a scattering process can be calculated in the same way as the Green function. The only difference is that now the external lines correspond to the wave functions rather than to the free Green functions.

Let us obtain the matrix elements for the scattering processes which we considered above,

$$
S = S^{(0)} + S^{(1)} + \cdots,
$$

where

In the zero order in γ (no interaction) everything is simple and for the sum of the two diagrams we have

$$
S^{(0)} = \delta(p_1 - p'_1)\delta(p_2 - p'_2) + \delta(p_1 - p'_2)\delta(p_2 - p'_1).
$$

The next order contribution is given by the processes with photon exchange:

Let us calculate the first diagram according to our graphical rules:

$$
S_a^{(1)}(y'_1, y'_2; x'_1, x'_2) = \delta(y'_1 - x'_1)\delta(y'_2 - x'_2)D(x'_2 - x'_1)\Gamma_\mu(x'_1)\Gamma_\mu(x'_2).
$$

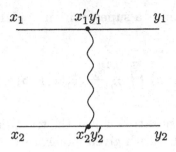

Then

$$S_a^{(1)}(p_1', p_2'; p_1, p_2) = \int d^4x_1' d^4x_2'\, D(x_2' - x_1')$$

$$\times \left(\frac{e^{ip_1'x_1'}}{\sqrt{2E_1'}} \gamma \frac{\overleftrightarrow{\partial}}{\partial x_{1\mu}'} \frac{e^{-ip_1 x_1'}}{\sqrt{2E_1}} \right) \left(\frac{e^{ip_2'x_2'}}{\sqrt{2E_2'}} \gamma \frac{\overleftrightarrow{\partial}}{\partial x_{2\mu}'} \frac{e^{-ip_2 x_2'}}{\sqrt{2E_2}} \right)$$

$$= \gamma^2 \int d^4x_1' d^4x_2'\, e^{i(p_1'-p_1)x_1'} e^{i(p_2'-p_2)x_2'} \int \frac{d^4k}{(2\pi)^4 i} \frac{e^{-ik(x_2'-x_1')}}{k^2}$$

$$\times \frac{[-i(p_1'+p_1)_\mu]}{\sqrt{2E_1 \cdot 2E_1'}} \frac{[-i(p_2'+p_2)_\mu]}{\sqrt{2E_2 \cdot 2E_2'}}$$

$$= \gamma^2 \int \frac{d^4k}{(2\pi)^4 i\, k^2} (2\pi)^4 \delta(p_1' - p_1 + k)(2\pi)^4 \delta(p_2' - p_2 - k)$$

$$\times \frac{[-i(p_1'+p_1)_\mu][-i(p_2'+p_2)_\mu]}{\sqrt{2E_1 \cdot 2E_2 \cdot 2E_1' \cdot 2E_2'}}$$

$$= \gamma^2 \frac{(2\pi)^4 i \delta(p_1' - p_1 + p_2' - p_2)}{\sqrt{2E_1 \cdot 2E_2 \cdot 2E_1' \cdot 2E_2'}} \left[\frac{(p_1 + p_1')_\mu (p_2 + p_2')_\mu}{(p_2' - p_2)^2} \right].$$

Similarly, we calculate the contribution $S_b^{(1)}$ of the second graph. The resulting expression is

$$S^{(1)} = (2\pi)^4 i \delta(p_1 + p_2 - p_1' - p_2') \frac{\gamma^2}{\sqrt{2E_1\, 2E_2\, 2E_1'\, 2E_2'}}$$

$$\times \left[\frac{(p_1 + p_1')_\mu (p_2 + p_2')_\mu}{(p_2 - p_2')^2} + \frac{(p_1 + p_2')_\mu (p_2 + p_1')_\mu}{(p_1' - p_2)^2} \right]. \tag{1.132}$$

Compare now (1.132) with (1.114) for the Green function. The difference between them is that, while in (1.114) free Green functions correspond to the external lines, in (1.132) it is the factor $1/\sqrt{2E}$ which corresponds to each external line. This is why it is convenient to pull out the factors $1/\sqrt{2E}$ from the matrix elements of the S-matrix.

1.11.3 Invariant scattering amplitude

Let us introduce the invariant scattering amplitude T via

$$S(p_1' \ldots p_n'; p_1, p_2) = 1 + (2\pi)^4 i\delta \left(p_1 + p_2 - \sum_{i=1}^{n} p_i' \right)$$
$$\times \frac{1}{\sqrt{2E_1}\sqrt{2E_2}} \prod_i \frac{1}{\sqrt{2E_i'}} T(p_1', \ldots, p_n'; p_1, p_2).$$

(1.133)

It is similar to the non-relativistic amplitude (1.37) we have derived above. For the sake of simplicity, we considered in (1.133) only the transition of two particles with momenta p_1, p_2 into n particles with momenta $p_1', \ldots, p_n'$. The probability of such a transition is

$$dW = \left[(2\pi)^4 \delta(p_1 + p_2 - \sum_{i=1}^{n} p_i') \right]^2 \frac{|T|^2}{2E_1\, 2E_2} \frac{d^3p_1' \ldots d^3p_n'}{(2\pi)^{3n}\, 2E_1' \ldots 2E_n'}. \quad (1.134)$$

To deal with the square of the delta-function we use the relations

$$[\delta(x)]^2 = \delta(x)\delta(0), \quad \text{and} \quad (2\pi)^4\delta(0) = \int d^4x\, e^{ipx}\,|_{p=0} = VT,$$

where V is the total volume of the three-dimensional space and T is the total time interval for the process. One of the δ-functions in (1.134) is cancelled in the transition probability *per unit volume and per unit time*, which thus becomes

$$dw \equiv \frac{dW}{VT} = (2\pi)^4\delta(p_1 + p_2 - \sum_{i=1}^{n} p_i') \frac{1}{2E_1 \cdot 2E_2}$$
$$\times |T(p_1', \ldots p_n'; p_1, p_2)|^2 \frac{d^3p_1' \ldots d^3p_n'}{(2\pi)^{3n} 2E_1' \ldots 2E_n'}.$$

(1.135)

1.11.4 Cross section

We are usually interested in the cross section $d\sigma$

$$d\sigma = \frac{dw}{j} = \frac{1}{4E_1 E_2 j}(2\pi)^4\delta\left(p_1 + p_2 - \sum_{i=1}^{n} p_i' \right)$$
$$\times |T(p_1', \ldots p_n'; p_1, p_2)|^2 \frac{d^3p_1' \ldots d^3p_n'}{(2\pi)^{3n}} \frac{1}{2E_1' \ldots 2E_n'},$$

(1.136)

where j is the flux of particles. The expression

$$\int \frac{d^3p'_1 \ldots d^3p'_n}{(2\pi)^{3n} \, 2E'_1 \ldots 2E'_n}$$

is called the invariant phase volume. (We met a similar expression when we calculated the Green functions.) Its relativistic invariance can be seen directly from comparison with (1.84) and (1.85).

For calculation of total cross sections the phase volume in (1.136) has to be divided by an additional factor $n!$ in the case of identical particles to avoid multiple counting of the identical configurations.

Let us determine the relative flux of the two colliding particles in the reference frame where their momenta are anti-collinear. Using the expression for the one-particle flux (the expression under the div operator on the right-hand side of (1.93)), we obtain

$$j = \frac{p_1}{E_1} - \frac{p_2}{E_2} = \frac{p_1 E_2 - E_1 p_2}{E_1 E_2}, \tag{1.137}$$

where p_1, p_2 are projections of the three-momenta on the collision axis. The numerator in (1.137) is invariant under boosts along the collision axis. It is called the invariant flux $\mathcal{J}$,

$$\mathcal{J} = 4E_1 E_2 \, j = 4(p_1 E_2 - E_1 p_2). \tag{1.138}$$

In the laboratory frame (say, $\mathbf{p}_2 = 0$) we have

$$\mathcal{J} = 4m p_L,$$

where p_L is the momentum of the projectile, and m is the mass of the particle. In the centre-of-mass frame, where $\mathbf{p}_1 = -\mathbf{p}_2$, $|\mathbf{p}_1| = |\mathbf{p}_2| \equiv p_c$,

$$\mathcal{J} = 4p_c E_c, \qquad E_c \equiv E_1 + E_2. \tag{1.139}$$

Hence, the cross section (1.136) of the process

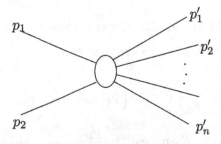

may be written in an explicitly Lorentz invariant form.

1.11.5 $2 \to 2$ scattering

Mandelstam variables. Let us now consider the case of $2 \to 2$ scattering

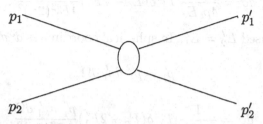

in detail. To describe such a process, it is convenient to introduce the invariant Mandelstam variables s, t and u:

$$\begin{aligned}
s &= (p_1 + p_2)^2 = (p_1' + p_2')^2 , \\
t &= (p_1' - p_1)^2 = (p_2 - p_2')^2 , \\
u &= (p_1 - p_2')^2 = (p_2 - p_1')^2 .
\end{aligned} \tag{1.140}$$

In the case of elastic scattering of particles with equal masses the Mandelstam variables have an especially simple interpretation in the centre-of-mass frame:

$$\begin{aligned}
s &= (p_{10} + p_{20})^2 = E_c^2 , \\
t &= -(\mathbf{p'}_1 - \mathbf{p}_1)^2 = -\mathbf{q}_c^2 , \\
u &= -(\mathbf{p}_1 - \mathbf{p}_2)^2 .
\end{aligned} \tag{1.141}$$

We see that s is the total energy squared in the c.m. frame, t is the momentum transfer squared between particles $1'$ and 1, and u is the momentum transfer squared between particles $2'$ and 1. The Mandelstam variables are not independent: they satisfy the relation

$$s + t + u = 4m^2 . \tag{1.142}$$

Indeed,

$$\begin{aligned}
s + t + u &= p_1^2 + p_2 + p_1'^2 + p_2'^2 + 2p_1^2 + 2p_1 p_2 - 2p_1 p_1' - 2p_1 p_2' \\
&= 4m^2 + 2p_1(p_1 + p_2 - p_1' - p_2') = 4m^2 .
\end{aligned}$$

For particles with different masses the corresponding relation reads $s + t + u = \sum_{i=1}^{4} m_i^2$.

Elastic scattering cross section. Now, with the help of (1.136), we are ready to write an expression for the elastic cross section. Let us go to the centre-of-mass frame where $\mathcal{J} = 4p_c E_c$ (see (1.139)). Then,

$$d\sigma = \frac{1}{4p_c E_c} |T|^2 (2\pi)^4 \delta(p_1 + p_2 - p_1' - p_2') \frac{d^3 p_1' \, d^3 p_2'}{4E_1' E_2' (2\pi)^6} .$$

Integrating over $\mathbf{p}'_2$, we get

$$d\sigma = \frac{1}{4p_c E_c}|T|^2 \delta(E_c - 2E'_1)\frac{d^3 p'_1}{4E'^2_1 (2\pi)^2},$$

where we have used $E'_2 = E'_1$. In spherical coordinates $d^3 p'_1$ can be written as

$$d^3 p'_1 = p'^2_1 \, dp'_1 \, d\Omega,$$

i.e.

$$d\sigma = \frac{1}{4p_c E_c}|T|^2 \delta(E_c - 2E'_1)\frac{p'^2_1 \, dp'_1 \, d\Omega}{4E^2_1 (2\pi)^2}.$$

Since $p'^2_1 = E'^2_1 - m^2$, we have $p'_1 dp'_1 = E'_1 dE'_1$. Hence,

$$\begin{aligned}
d\sigma &= \frac{1}{4p_c E_c}|T|^2 \delta(E_c - 2E'_1)\frac{p_c \, E'_1 dE'_1 d\Omega}{4E'^2_1 (2\pi)^2} \\
&= \frac{1}{4E}|T|^2 \delta(E_c - 2E'_1)\frac{dE'_1 d\Omega}{4E'_1 (2\pi)^2}.
\end{aligned}$$

After integration over the energy E'_1, and using the relations $4E'_1 = 2E_c$ and $\int dE'_1 \delta(E_c - 2E'_1) = 1/2$, we finally obtain

$$d\sigma = \frac{1}{16E^2_c}\frac{d\Omega}{(2\pi)^2}|T|^2. \tag{1.143}$$

In terms of the Mandelstam variables (1.143) is

$$d\sigma = \left|\frac{T}{8\pi\sqrt{s}}\right|^2 d\Omega. \tag{1.144}$$

1.11.6 $\pi^- \pi^-$ scattering

Let us now return to the Coulomb scattering $\pi^- \pi^-$. In the lowest order this process is described by the diagrams

Comparing (1.133) and (1.132), it is easy to write the invariant scattering amplitude

$$T(p'_2, p'_1; p_2, p_1) = \gamma^2 \left[\frac{(p_1 + p'_1)_\mu (p_2 + p'_2)_\mu}{(p'_1 - p_1)^2} + \frac{(p_1 + p'_2)_\mu (p_2 + p'_1)_\mu}{(p'_1 - p_2)^2}\right]. \tag{1.145}$$

First, we express the numerators in the brackets in terms of invariant variables. Calculation of the first numerator can be simplified by introducing the u-channel momentum transfer $r = p_1 - p'_2 = -p_2 + p'_1$, $r^2 = u$. Then we have

$$(p_1 + p'_1)_\mu (p_2 + p'_2)_\mu = (p_1 + p_2 + r)_\mu (p_1 + p_2 - r)_\mu = s - u.$$

Similarly, the second numerator gives $s - t$. As a result,

$$T = \gamma^2 \left[\frac{s-u}{t} + \frac{s-t}{u} \right]. \tag{1.146}$$

Expression (1.146) contains only one unknown constant γ^2, which can be determined from scattering experiments. However, there is no need to carry out experiments for this purpose, since in the region of small momenta (1.146) should coincide with the well-known non-relativistic formula for Coulomb scattering.

Non-relativistic limit. To obtain the non-relativistic approximation we consider again the centre-of-mass reference frame, where

$$\mathbf{p}_1 = -\mathbf{p}_2 ; \qquad \mathbf{p}'_1 = -\mathbf{p}'_2,$$

$$p_{10} = p_{20} = p'_{10} = p'_{20},$$

and

$$
\begin{aligned}
s &= (p_{10} + p_{20})^2 - (\mathbf{p}_1 + \mathbf{p}_2)^2 = (p_{10} + p_{20})^2 = E_c^2 ; \\
t &= (p'_{10} - p_{10})^2 - (\mathbf{p}_1 - \mathbf{p}'_1)^2 = -|\mathbf{p}_1|^2 - |\mathbf{p}'_1|^2 + 2|\mathbf{p}_1||\mathbf{p}'_1| \cos\theta \\
&= -2p_c^2(1 - \cos\theta) = -\mathbf{q}^2, \qquad \mathbf{q} = \mathbf{p}'_1 - \mathbf{p}_1 ; \\
u &= -2p_c^2(1 + \cos\theta) = -\mathbf{q}'^2, \qquad \mathbf{q}' = \mathbf{p}'_2 - \mathbf{p}_1 .
\end{aligned}
$$

In the non-relativistic limit $s = E_c^2 \simeq 4m^2$, the momentum transfer invariants are relatively small, $|t|$, $|u| \ll s$, and (1.146) becomes

$$T \simeq -\gamma^2 4m^2 \left[\frac{1}{\mathbf{q}^2} + \frac{1}{\mathbf{q}'^2} \right]. \tag{1.147}$$

On the other hand, in non-relativistic quantum mechanics the scattering amplitude f in the Born approximation has the form (compare with (1.38))

$$f = -\frac{2\mu}{4\pi} \int e^{-i\mathbf{q}\mathbf{r}} U(\mathbf{r}) d^3 r + f_{\text{exchange}} ,$$

where $\mu = m/2$ is the reduced mass. In the Heaviside units, the Coulomb potential has the form

$$U(r) = \frac{e^2}{4\pi r} ,$$

and consequently,

$$f = -\frac{m}{4\pi}\frac{e^2}{\mathbf{q}^2} + f_{\text{exchange}}. \tag{1.148}$$

The exchange terms in (1.147) and (1.148) can be neglected for small scattering angles $\theta \ll 1$.

Now we are in a position to establish the connection between T and the non-relativistic amplitude f. In terms of f, the cross section is

$$d\sigma = |f|^2 d\Omega.$$

Comparing this expression with (1.144) we establish the relative normalization of the amplitudes,

$$f = \frac{T}{8\pi\sqrt{s}} \simeq \frac{T}{16\pi m}. \tag{1.149}$$

Inserting the values f (1.148) and T (1.147) into (1.149), we can now determine the constant γ^2:

$$-\frac{m}{2\pi}e^2\frac{1}{\mathbf{q}^2} = -\gamma^2\frac{4m^2}{\mathbf{q}^2 16\pi m},$$

which leads to

$$\gamma^2 = e^2. \tag{1.150}$$

In our units $e^2/4\pi = 1/137 \ll 1$. Thus, indeed, with high accuracy it is sufficient to consider only the simplest processes.

$\pi^-\pi^-$ *invariant scattering amplitude.* Substituting $\gamma^2 = e^2$ in (1.146), we obtain the scattering amplitude T for arbitrary energies. In the centre-of-mass frame it takes the form

$$T = -e^2 \left[\frac{3 + \cos\theta + \frac{2m^2}{\mathbf{p}^2}}{1 - \cos\theta} + \frac{3 - \cos\theta + \frac{2m^2}{\mathbf{p}^2}}{1 + \cos\theta} \right]. \tag{1.151}$$

Note that in the ultra-relativistic limit, $|\mathbf{p}| \gg m$, the amplitude depends only on the centre-of-mass scattering angle and not on the momentum of the projectile. The angular dependence of T is as follows:

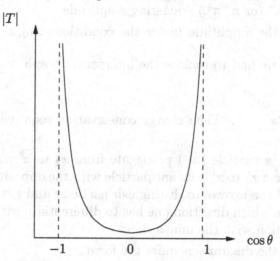

To summarize, we have shown that the coupling constant for the interaction of charged spinless particles with the electromagnetic field coincides with the charge of these particles, i.e. $\gamma^2 = e^2$. In addition, we have obtained the expression (1.151) for the invariant $\pi^-\pi^-$ (or $\pi^+\pi^+$) scattering amplitude in the first order in e^2.

1.11.7 $\pi^+\pi^-$ scattering

Now we turn to $\pi^+\pi^-$ scattering:

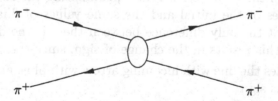

Let us recall the logic which led us to introducing π^+ as an antiparticle to π^-. In the coordinate representation, the lowest order diagrams describing scattering of identical charged particles were

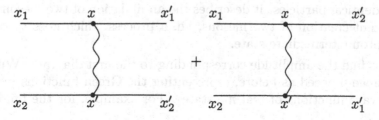

In the limit $x_{10}, x_{20} \to -\infty$, $x'_{10}, x'_{20} \to +\infty$ we derived from these diagrams the $\pi^-\pi^-$ (or $\pi^+\pi^+$) scattering amplitude.

Calculating the amplitude under the conditions $x_{10}, x'_{20} \to -\infty$ and $x_{20}, x'_{10} \to \infty$, we had to replace the interaction graph

by $\qquad$. Then charge conservation required the following interpretation: a particle (π^-) propagate from x_2 to x' while the object propagating from x'_2 to x' is its antiparticle with the opposite charge (π^+). We introduced the arrows to distinguish particles and antiparticles, and to remember in which direction one has to differentiate with the plus and in which direction with the minus sign.

As a result, the diagrams acquire the form

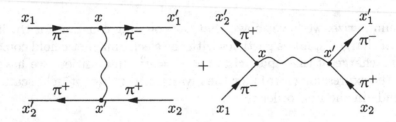

and describe $\pi^-\pi^+$ scattering. Two comments are in order. The first of the two diagrams describing $\pi^-\pi^+$ scattering differs in sign from the first diagram in the $\pi^-\pi^-$ case. Indeed, comparing the two graphs (with the same values of the initial and the same values of final coordinates) we observe that the only difference between them is the direction of the bottom line. This results in the change of sign, since the vertex operator $\overleftrightarrow{\partial}_\mu$ differentiates the line with incoming arrow with plus, and the outgoing one with minus sign.

We also remark that because π^- and π^+ are not identical objects, there is no diagram with a simple interchange of the final particles as in the $\pi^-\pi^-$ case. We have instead the second $\pi^-\pi^+$ graph which is essentially different from those corresponding to the virtual photon exchange between identical particles: it describes the annihilation of two mesons with subsequent creation of two mesons, i.e. a process which goes through a one-photon intermediate state.

Let us find the amplitude corresponding to the first diagram. With the top line we proceed as before, representing the Green functions in terms of the wave functions of real π^- states. For example, for the final state

π^- we write

$$G(x_1' - x) = \int \frac{d^4 p_1'}{(2\pi)^4 i} \frac{e^{-ip_1'(x_1'-x)}}{m^2 - p_1'^2} = \int \frac{d^3 p_1'}{(2\pi)^3} \, \Psi_{p_1'}(x_1') \, \Psi_{p_1'}^*(x) \, . \quad (1.152)$$

We can proceed similarly with the bottom line. For the outgoing π^+, in particular, we can write

$$G(x_2 - x') = \int \frac{d^3 p_2^+}{(2\pi)^3} \, \Psi_{p_2^+}(x_2) \, \Psi_{p_2^+}^*(x') \, , \quad (1.153)$$

where

$$\Psi_{p_2^+}(x) = \frac{e^{-ip_2^+ x}}{\sqrt{2p_0}}, \qquad p_{20}^+ = \sqrt{m^2 + \mathbf{p}_2^{+2}} \, .$$

We do the same for all external particles.

Since the direction of the bottom line is reversed, the photon emission vertex expressed in terms of π^+ momenta, $\propto (p_2^+ + p_2^{+\prime})$, has the opposite sign as compared with the corresponding expression describing the top π^- line, $\propto (p_1' + p_1)$. The necessity to keep in mind the sign of the differentiation is rather inconvenient. Instead, one can write the amplitude for the $\pi^- \pi^-$ scattering, and close the loop around the pole in the corresponding Green function in accordance with the conditions $x_{10}, x_{20}' \to -\infty$, $x_{20}, x_{10}' \to +\infty$. In this case we have for the bottom line Green function

$$G(x' - x_2) = \int \frac{d^4 p_2}{(2\pi)^4 i} \frac{e^{-ip_2(x'-x_2)}}{m^2 - p_2^2 - i\varepsilon},$$

except that here the contour has to be closed around the negative-energy pole $p_{20} = -\sqrt{\mathbf{p}_2^2 + m^2}$, in the upper half-plane. Then,

$$\begin{aligned}
G(x' - x_2) &= \int \frac{d^3 p_2}{(2\pi)^3 2 \, |p_{20}|} \, e^{i\sqrt{\mathbf{p}_2^2 + m^2}(x_0' - x_{20}) + i p_2 (\mathbf{x}' - \mathbf{x}_2)} \\
&= \int \frac{d^3 p_2}{(2\pi)^3} \, \Psi_{-p_2}(x_2) \, \Psi_{-p_2}^*(x') ,
\end{aligned} \quad (1.154)$$

where we used that the substitution of $\mathbf{p}_2$ by $-\mathbf{p}_2$ does not change the result. Here

$$\Psi_{-p_2}(x) = \frac{e^{ip_2 x}}{\sqrt{2|p_{20}|}}, \qquad p_{20} = -\sqrt{\mathbf{p}_2^2 + m^2} \, .$$

The expressions (1.154) and (1.153) are identical and describe the propagation of a positive-energy π^+ with four-momentum $p_2^+ = -p_2$ from x' to x_2 ($x_0' < x_{20}$). There is, however, an alternative way to represent (1.154).

Observing that permutation of coordinates is equivalent to changing sign of the four-momentum, we may write

$$G(x' - x_2) = \int \frac{d^3 p_2}{(2\pi)^3} \, \Psi_{-p_2}(x_2) \, \Psi^*_{-p_2}(x') = \int \frac{d^3 p_2}{(2\pi)^3} \, \Psi_{p_2}(x') \, \Psi^*_{p_2}(x_2) \, .$$

We see that the propagation of the π^+-meson from x' to x_2 can be described either in terms of $\Psi_{p_2^+}$ where $p_2^+ = -p_2$ is a momentum corresponding to positive energy, or in terms of the wave function Ψ_{p_2} of the π^--meson with a negative energy, $(p_{20} < 0)$, propagating from x_2 to x'. This observation leads to the Feynman interpretation of an antiparticle as a negative energy particle which propagates backwards in time.

Given that the substitution $p_2^+ = -p_2$ turns the propagator of the π^--meson moving from x_2 to x' into the propagator of the π^+-meson which moves from x' to x_2, we can get the $\pi^+\pi^-$ scattering amplitude from the $\pi^-\pi^-$ amplitude simply via substitutions of momenta.

Consider the process $\pi^-(p_1) + \pi^+(p_2^+) \rightarrow \pi^-(p_1') + \pi^+(p_2'^+)$, which, as we know, is described by the sum of the scattering and annihilation diagrams (see page 72),

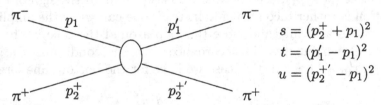

$$s = (p_2^+ + p_1)^2$$
$$t = (p_1' - p_1)^2$$
$$u = (p_2^{+'} - p_1)^2$$

To obtain the corresponding amplitude we take the invariant $\pi^-\pi^-$ scattering amplitude (1.145),

$$T(p_2', p_1'; p_2, p_1) = e^2 \left[\frac{(p_1 + p_1')_\mu (p_2 + p_2')_\mu}{(p_1 - p_1')^2} + \frac{(p_1 + p_2')_\mu (p_2 + p_1')_\mu}{(p_1' - p_2)^2} \right] ,$$

and make the substitution

$$p_2^+ = -p_2' , \quad p_2'^+ = -p_2 , \tag{1.155}$$

which turns one of the initial (final) π^--mesons into the final (initial) π^+. This gives us the scattering amplitude for $\pi^-\pi^+$:

$$T(p_2'^+, p_1'; p_2^+, p_1)$$
$$= e^2 \left[-\frac{(p_1 + p_1')_\mu (p_2^+ + p_2'^+)_\mu}{(p_1 - p_1')^2} + \frac{(p_1 - p_2'^+)_\mu (p_1' - p_2^+)_\mu}{(p_1' + p_2^+)^2} \right] .$$

A calculation similar to (1.146) leads to the following expression in terms of the Mandelstam variables s, t, u:

$$T_{\pi^+\pi^-} = e^2 \left[-\frac{s - u}{t} + \frac{u - t}{s} \right] . \tag{1.156}$$

Actually, we do not even need to perform this calculation from scratch. All we need to do is to find how s, t and u defined in (1.140) are transformed under (1.155):

$$s_{\pi^-\pi^-} \to u_{\pi^-\pi^+}, \quad t_{\pi^-\pi^-} \to t_{\pi^-\pi^+}, \quad u_{\pi^-\pi^-} \to s_{\pi^+\pi^-}.$$

Next we just substitute the transformed variables into the final expression (1.146) for the invariant scattering amplitude describing $\pi^-\pi^-$ scattering and, lo and behold, what we have really obtained is (1.156).

We will study amplitudes of the type (1.156) in the case of electron scattering in more detail in Section 2.5.

1.12 The Mandelstam plane

Now let us discuss the connection between amplitudes in a more general way for the case of two charged particles, for example the elastic scattering $\pi^-\pi^- \to \pi^-\pi^-$.

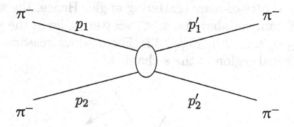

As we mentioned before, the Mandelstam variables are Lorentz invariant and satisfy the condition

$$s + t + u = 4m^2. \tag{1.157}$$

This relation can easily be visualised with the help of the Mandelstam plane in Fig. 1.12 where each point corresponds to given values of the Mandelstam variables s, t, u satisfying (1.157). Here we use the fact that the sum of the altitudes of an equilateral triangle does not depend on the position of a point. (The extended sides of an equilateral triangle play the role of the coordinate axes, and the altitudes are counted with sign; the arrows in Fig. 1.12 mark the positive directions.)

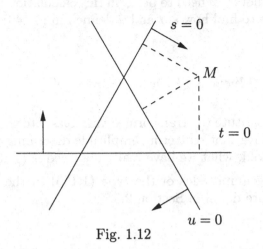

Fig. 1.12

Let us find the physical region of the reaction $\pi^-\pi^-$ in the Mandelstam plane. In the centre-of-mass frame we have $p_c = |\mathbf{p_1}| = |\mathbf{p_2}| = |\mathbf{p_1'}| = |\mathbf{p_2'}|$ and therefore

$$s = 4(m^2 + p_c^2) \geq 4m^2\,,$$
$$t = -2p_c^2(1 - \cos\theta) \leq 0\,,$$
$$u = -2p_c^2(1 + \cos\theta) \leq 0\,,$$

where θ is the centre-of-mass scattering angle. Hence, the allowed values of the Mandelstam variables for $\pi^-\pi^-$ scattering lie in the shaded region $s \geq 4m^2$, $t \leq 0$, $u \leq 0$ in Fig. 1.13. For obvious reasons this region is called the physical region of the s-channel.

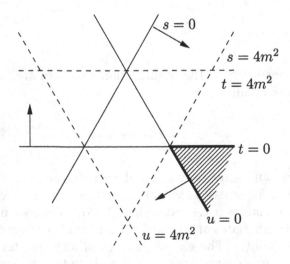

Fig. 1.13

The bold lines in Fig. 1.13 mark the small-t region corresponding to forward scattering and the small-u region which corresponds to backward scattering. (Also shown in Fig. 1.13 are the straight lines passing parallel to the sides of the triangle through its vertices with the coordinates $(s = 4m^2, t = u = 0)$, $(t = 4m^2, s = u = 0)$, and $(u = 4m^2, s = t = 0)$.)

In general one has three non-overlapping regions in the Mandelstam plane which correspond to three different channels as shown in Fig. 1.14. The shaded region $u \geq 4m^2$, $s \leq 0$, $t \leq 0$ is called the u-channel, and the shaded region $t \geq 4m^2$, $u \leq 0$, $s \leq 0$ is called the t-channel.

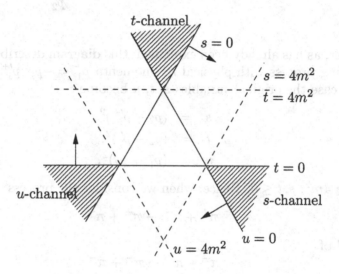

Fig. 1.14

Consider, for example, what physical process corresponds to the u-channel region. Make the substitution $p'_1 = -p_1^+$; $p_1 = -p_1'^+$ in the amplitude of the process

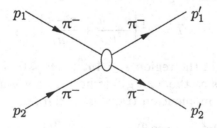

Here $p_0^+ > 0$, i.e. we analytically continue the original amplitude into the negative-frequency region. The diagram can then be represented as

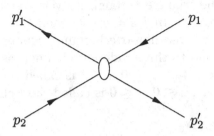

However, as has already been explained, this diagram describes the $\pi^+\pi^-$ scattering process with physical π^+-momenta $p_1^+ = -p_1'$, $p_1'^+ = -p_1$, and in this case the original variables s, t, u become

$$
\begin{aligned}
s &= (p_2 - p_1'^+)^2, \\
t &= (p_2 - p_2')^2, \\
u &= (p_1^+ + p_2)^2,
\end{aligned}
$$

i.e. $u \geq 4m^2$; $s, t \leq 0$. Hence, when we consider the process

$$\pi^+ + \pi^- \rightarrow \pi^+ + \pi^-$$

instead of

$$\pi^- + \pi^- \rightarrow \pi^- + \pi^-,$$

our Mandelstam variables change, and we go from the s-channel region to the u-channel region. By doing so, we have in fact analytically continued the scattering amplitude T from the s-channel to the u-channel, where it describes a different process. This is the essence of how we obtained (1.156) from (1.146).

Indeed, our original amplitude was

$$T = e^2 \left[\frac{s-u}{t} + \frac{s-t}{u} \right],$$

and it was defined in the region $s \geq 4m^2$; $t, u \leq 0$. If now we look upon $u = 4(p_c^2 + m^2) \equiv \tilde{s}$ as the 'energy' (whose rôle was played by s before), and t and s as the 'momentum transfer' variables,

$$t = -2p_c^2(1 - \cos\theta), \quad s = -2p_c^2(1 + \cos\theta) \equiv \tilde{u},$$

then

$$T_{\pi^+\pi^-} = e^2 \left[\frac{\tilde{u} - \tilde{s}}{t} + \frac{\tilde{u} - t}{\tilde{s}} \right].$$

This is just the expression for the $\pi^+\pi^-$ scattering amplitude we had earlier.

Thus, there are two ways of connecting the amplitudes for scattering of particles and of antiparticles. Either we calculate the amplitudes and consider different positive-frequency and negative-frequency momenta, or we fix the variables s, t and u in a definite process and consider their different values afterwards.

Let us discuss one more example:

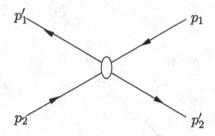

Looking at the graph 'from the top', we have p_1 and p_2' representing particles, and $p_1' = -p_1^{+\prime}$ and $p_2 = -p_2^+$ antiparticles. From this point of view,

$$
\begin{aligned}
s &= (p_1 - p_2^+)^2, \\
u &= (p_1 - p_2')^2, \\
t &= (p_1 + p_1^{+\prime})^2,
\end{aligned}
$$

i.e. we are now in the physical region of the t-channel, which corresponds to the reaction $\pi^+(p_1^{+\prime}) + \pi^-(p_1) \to \pi^+(p_2^+) + \pi^-(p_2')$.

Hence, we obtained the following important result: an amplitude describes not one process, but a whole class of processes. Namely,

$$\pi^- + \pi^- \to \pi^- + \pi^- \qquad \text{in the } s\text{-channel};$$

$$\pi^+ + \pi^- \to \pi^+ + \pi^- \qquad \text{in the } u\text{-channel};$$

$$\pi^+ + \pi^- \to \pi^- + \pi^+ \qquad \text{in the } t\text{-channel}.$$

Finally, note that the decay of the π-meson into three π-mesons is forbidden by energy–momentum conservation, i.e. it lies in the non-physical region. However, if we increase the mass of one of the particles, the same amplitude will also describe the decay process. The physical region of the decay process is located inside the triangle in the Mandelstam plane.

1.13 The Compton effect (for π-mesons)

The simplest diagrams which describe photon scattering off the π-meson are

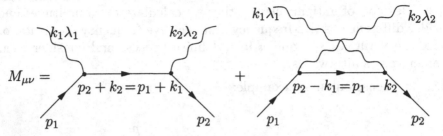

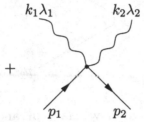

The wave function of the meson is

$$\frac{e^{-ipx}}{\sqrt{2p_0}},$$

that of the photon

$$e_\mu^\lambda(k)\frac{e^{-ikx}}{\sqrt{2k_0}}.$$

We have obtained these from the transition amplitudes in the limit $t \to \infty$. Further, we have for the photon

$$D_{\mu\nu}(x) \;=\; g_{\mu\nu}\int\frac{d^4k}{(2\pi)^4 i}\frac{e^{-ikx}}{k^2} = (-g_{\mu\nu})\int\frac{d^3k}{(2\pi)^3}\frac{e^{-ikx}}{2k_0},$$

$$-g_{\mu\nu} \;=\; \sum_{\lambda=0}^{3} e_\mu^\lambda e_\nu^{\lambda*}.$$

The polarization vectors e_μ^λ and e_ν^λ were associated with different wave functions. Similarly to the case of $\pi\pi$ scattering, the amplitudes corresponding to the first two diagrams can be written immediately. Denoting by $\tilde{M}_{\lambda_2\lambda_1}$ the amplitude which corresponds to the third graph, we obtain

$$T_{\pi\gamma} = (p_2 + p_2 + k_2)_\nu e_\nu^{\lambda_2*}\frac{e^2}{m^2 - (p_2 + k_2)^2}(p_1 + p_1 + k_1)_\mu e_\mu^{\lambda_1}$$

$$+ (p_2 + p_2 - k_1)_\mu e_\mu^{\lambda_1}\frac{e^2}{m^2 - (p_2 - k_1)^2}(p_1 + p_1 - k_2)_\nu e_\nu^{\lambda_2*} + \tilde{M}_{\lambda_2\lambda_1}.$$

$$(1.158)$$

Let us now extract the factor $e_\nu^{\lambda_2 *} e_\mu^{\lambda_1}$ from (1.158) and call the remaining tensor $M_{\nu\mu}$, i.e.

$$T_{\pi\gamma} = e_\nu^{\lambda_2 *} e_\mu^{\lambda_1} \, M_{\nu\mu}. \qquad (1.159)$$

As shown in Section 1.8, the conditions

$$k_{1\mu} M_{\nu\mu} = 0, \quad k_{2\nu} M_{\nu\mu} = 0 \qquad (1.160)$$

have to be satisfied to avoid the production of longitudinally polarized photons. For $k^2 \neq 0$, there are three vectors orthogonal to k_μ: $e_\mu^\lambda k_\mu = 0$ ($\lambda = 1, 2, 3$), namely:

$$e^{(1)} = \begin{pmatrix} 0 \\ 1 \\ 0 \\ 0 \end{pmatrix}, \quad e^{(2)} = \begin{pmatrix} 0 \\ 0 \\ 1 \\ 0 \end{pmatrix}, \quad e^{(3)} = \frac{1}{\sqrt{k^2}} \begin{pmatrix} |\mathbf{k}| \\ 0 \\ 0 \\ k_0 \end{pmatrix}.$$

For $\lambda = 0$ we have $e_\mu^{(0)} \propto k_\mu$ (see Section 1.3). We here use the notation

$$e = \begin{pmatrix} e_0 \\ e_x \\ e_y \\ e_z \end{pmatrix},$$

with z the direction of the photon momentum $\mathbf{k}$. Since $e_\mu^{(0)}$ is proportional to k_μ, the terms with the scalar polarization vector $e^{(0)}$ do not make any contribution in (1.159) if (1.160) is satisfied. For real photons ($k^2 = 0$) the terms with the longitudinal polarization $e^{(3)}$ also vanish. Thus, scalar and longitudinal photons do not contribute to the physical processes if the condition (1.160) is fulfilled.

The tensor $M_{\nu\mu}$ has the following form:

$$M_{\nu\mu} = e^2 \left[\frac{(2p_2 + k_2)_\nu (2p_1 + k_1)_\mu}{-2\,p_2 k_2} + \frac{(2p_2 - k_1)_\mu (2p_1 - k_2)_\nu}{2\,p_2 k_1} \right] + \tilde{M}_{\nu\mu}.$$

Let us calculate $k_{2\nu} M_{\nu\mu}$ taking into account that $p_1 + k_1 = p_2 + k_2$ and, hence, $2p_2 k_1 = 2p_1 k_2$:

$$k_{2\nu} M_{\nu\mu} = e^2 [-(2p_1 + k_1)_\mu + (2p_2 - k_1)_\mu] + k_{2\nu} \tilde{M}_{\nu\mu}.$$

Since $p_2 - p_1 - k_1 = -k_2$, we arrive at

$$k_{2\nu} M_{\nu\mu} = -2e^2 k_{2\mu} + k_{2\nu} \tilde{M}_{\nu\mu}.$$

We conclude that (1.160) can be satisfied, i.e. the current can be conserved, only if the term corresponding to contact interactions is introduced.

The simplest guess for the contact interaction is to take

$$\tilde{M}_{\nu\mu} = 2\,e^2\,g_{\nu\mu}\,. \qquad (1.161)$$

This choice gives us an amplitude for which the current is conserved and which does not allow scalar and longitudinal photons to participate in the interaction.

We can now calculate $T_{\pi\gamma}$ using the contact interaction introduced in (1.161). Taking advantage of relativistic invariance of the amplitude (1.158), we simplify the calculation by working in the rest frame of the initial electron:

$$p_{10} = m\,, \quad \mathbf{p_1} = 0\,.$$

We have to calculate the amplitude (1.158) for the two physical polarization vectors orthogonal to k_μ. Hence, all terms in (1.158) which are proportional to $k_{1\mu}$ or $k_{2\nu}$ do not contribute. Moreover, since these two physical polarization vectors have only space components and the vector p_1 has only a time component in our reference frame, the first two terms on the right-hand side of (1.158) vanish, and only the contact (seagull) term gives a non-zero contribution to the amplitude. Thus,

$$T_{\pi\gamma} = e_\nu^{\lambda_2*} e_\mu^{\lambda_1} g_{\mu\nu} \cdot 2e^2\,,$$

i.e.

$$T_{\pi\gamma} = 2e^2 \left(e^{\lambda_2*}(\mathbf{k_2})\, e^{\lambda_1}(\mathbf{k_1}) \right)\,.$$

In the case of small-angle scattering, $\mathbf{k_1}/|\mathbf{k_1}| \simeq \mathbf{k_2}/|\mathbf{k_2}|$, we simply obtain

$$T_{\pi\gamma} = -2e^2\,\delta_{\lambda_1,\lambda_2}\,.$$

Let us investigate the connection with the usual non-relativistic scattering amplitude. As we have seen above in (1.149),

$$f = \frac{T}{8\pi\sqrt{s}} = -\frac{e^2}{4\pi\sqrt{s}}\,. \qquad (1.162)$$

In the limit of small photon frequency, $k \to 0$, we have $\sqrt{s} \to m$, i.e.

$$f \simeq -\frac{e^2}{4\pi m} = -\frac{e'^2}{m}\,,$$

which coincides with the expression for the classical non-relativistic Thomson scattering amplitude, with e' the usual (non-Heaviside) charge.

Now consider our amplitude from the point of view of different channels. The replacement

$$k_2 = -k_1^+,$$
$$k_1 = -k_2^+$$

means that we go to the u-channel. Such a substitution interchanges only the γ quanta, and, since they are neutral, the amplitude in the u-channel turns out to be the same as in the s-channel. The substitution

$$
\begin{aligned}
p_2 &= -p_2^+, \\
k_1 &= -k_1^+
\end{aligned}
$$

leads to a new process, namely, to the two-photon annihilation of two mesons π^+ and π^-. It corresponds to the transition into the t-channel region.

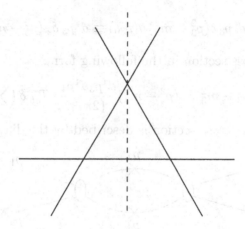

The symmetry of the amplitude with respect to the dashed line in the Mandelstam plane reflects the neutrality of the photon.

p_1 $k_1^+ = -k_1$

$p_2^+ = -p_2$ k_2

So far, we have learned how to calculate the amplitudes of different processes with the help of the Feynman diagrams. Let us see whether we can directly calculate *cross sections* from the diagrams. We have already written the cross section for the process

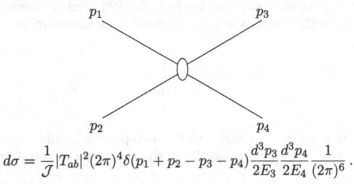

$$d\sigma = \frac{1}{\mathcal{J}}|T_{ab}|^2 (2\pi)^4 \delta(p_1 + p_2 - p_3 - p_4)\frac{d^3p_3}{2E_3}\frac{d^3p_4}{2E_4}\frac{1}{(2\pi)^6}\,.$$

Using the simple relation which is valid for the integrands,

$$\frac{d^3p_3}{2E_3} = d^4p_3\,\delta(p_3^2 - m_3^2)\,\theta(p_{30}) \equiv d^4p_3\,\delta_+(p_3^2 - m_3^2)\,,$$

we can cast the cross section in the following form:

$$d\sigma = \frac{1}{\mathcal{J}}T_{ab}\cdot\delta_+(p_3^2 - m_3^2)\delta_+(p_4^2 - m_4^2)\frac{d^4p_3 d^4p_4}{(2\pi)^6}\cdot T_{ab}^*\,\delta\left(\sum p_i\right). \quad (1.163)$$

This means that the cross section is described by the diagram

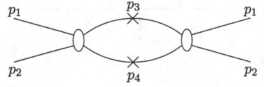

which has to be calculated by our usual rules except that the lines with crosses correspond now to the δ-functions instead of the Green functions. This is quite natural because the intermediate states in the case of the cross section correspond to real particles for which $p_{30} = \sqrt{m_3^2 + \mathbf{p}_3^2}$ and $p_{40} = \sqrt{m_4^2 + \mathbf{p}_4^2}$.

Note in passing that this result could also be derived from the representation of the Green function in the form

$$G = \frac{1}{m^2 - p^2 - i\varepsilon} = P\frac{1}{m^2 - p^2} + i\pi\delta(p^2 - m^2)\,. \quad (1.164)$$

Here the symbol P stands for the integration in the sense of the principal value.

The form (1.163) is convenient since it allows us to continue the cross section from one channel to the other (provided that during this process the amplitude T will not acquire an imaginary part). We will discuss this analytic continuation later in Section 3.3.

2

Particles with spin $\frac{1}{2}$. Basic quantum electrodynamic processes

2.1 Free particles with spin $\frac{1}{2}$

A spin $J = \frac{1}{2}$ particle can be described by two probability amplitudes $\varphi_\lambda(\lambda = 1, 2)$ of finding a particle in a state with spin projections $\frac{1}{2}$ and $-\frac{1}{2}$, respectively. They can be written in the form $\binom{\varphi_1}{\varphi_2}$. Hence, the wave function of a particle at rest is, as usual,

$$\Psi_0 = \binom{\varphi_1}{\varphi_2} e^{-imt}. \tag{2.1}$$

Recall that the wave function for a moving $J = 0$ particle was simply

$$\Psi = \frac{e^{-ipx}}{\sqrt{2p_0}}.$$

In the present case the situation is more complicated. To obtain the wave function of a particle with finite momentum, one has to carry out a transformation from the rest frame to a moving reference frame. To do this we need to know how φ_λ changes under the Lorentz transformation. In other words, we have to find a representation of the Lorentz group acting on our two-component objects.

Lorentz transformations have the form

$$x_i' = a_{ik} x_k, \tag{2.2}$$

where the transformation matrix depends on six parameters: three Euler angles θ_i which parametrize spatial rotations and three components of the relative velocity vector v_i describing the transition from one inertial reference frame to another.

Now consider the wave functions. Generally speaking, the components of the wave function ξ_1 and ξ_2 transform as

$$\xi_1' = u_{11}\xi_1 + u_{12}\xi_2,$$

$$\xi'_2 = u_{21}\xi_1 + u_{22}\xi_2. \tag{2.3}$$

Since both ξ_i and u_{ik} are complex, the matrix u_{ik} contains eight independent parameters.

It is easy to show that matrices with unit determinant

$$\det(u_{ik}) = u_{11}u_{22} - u_{12}u_{21} = 1, \tag{2.4}$$

also form a group. Equation (2.4) gives two independent conditions for the real and imaginary parts, and, hence, the matrix u_{ik} is characterized by six independent parameters. These parameters can be connected with the parameters of Lorentz transformations. Thus, complex two-dimensional matrices with unit determinant realize a representation of the Lorentz group.

As we know, any 2×2 matrix can be written as a linear combination of four matrices (a unit matrix and three Pauli matrices):

$$I = \begin{pmatrix} 1 & 0 \\ 0 & 1 \end{pmatrix}, \ \sigma_x = \begin{pmatrix} 0 & 1 \\ 1 & 0 \end{pmatrix}, \ \sigma_y = \begin{pmatrix} 0 & -i \\ i & 0 \end{pmatrix}, \ \sigma_z = \begin{pmatrix} 1 & 0 \\ 0 & -1 \end{pmatrix}.$$

The Pauli matrices have the following properties:

$$[\sigma_i \sigma_k] = 2i\epsilon_{ikl}\sigma_l,$$
$$\sigma_x\sigma_y = i\sigma_z, \quad \sigma_y\sigma_z = i\sigma_x, \quad \sigma_z\sigma_x = i\sigma_y. \tag{2.5}$$

Note that the matrices $\sigma_i/2$ have the same commutation properties as the rotations.

We will parametrize rotations in the usual three-dimensional space by the vector $\boldsymbol{\theta}$ which is directed along the rotation axis and has a length equal to the magnitude of the rotation angle. Under three-dimensional rotations the wave function $\xi = \begin{pmatrix} \xi_1 \\ \xi_2 \end{pmatrix}$ transforms as

$$\xi' = u\,\xi, \tag{2.6}$$

where

$$u = e^{\frac{i}{2}\sigma\cdot\theta} = e^{\frac{i}{2}\sigma_n\theta_n}. \tag{2.7}$$

The complex conjugate wave function ξ^* transforms as

$$\xi'^* = e^{-\frac{i}{2}\sigma_n^*\theta_n}\xi^*.$$

Transposing this relation, we get

$$\xi'^{*\mathrm{T}} = \xi^{*\mathrm{T}}e^{-\frac{i}{2}\sigma_n^{*\mathrm{T}}\theta_n},$$

or, due to the hermiticity of the Pauli matrices,

$$\xi'^\dagger = \xi^\dagger e^{-\frac{i}{2}\sigma_n\theta_n}. \tag{2.8}$$

Here the symbol $*$ denotes complex conjugation, $\top$ stands for transposition, $\dagger$ denotes Hermitian conjugation, and $\xi^\dagger$ is a row (ξ_1^*, ξ_2^*). The product

$$(\xi^\dagger\xi) = \xi_1^*\xi_1 + \xi_2^*\xi_2 \tag{2.9}$$

is the usual scalar product in two-dimensional space. Using (2.7) and (2.8), it is easy to see that

$$(\xi'^\dagger\xi') = (\xi^\dagger\xi). \tag{2.10}$$

Thus u-matrices form a unitary two-dimensional representation of three-dimensional rotations.

Now consider the transformation law of the three-component object

$$A_i = \xi^\dagger\sigma_i\xi,$$

which after the transformation becomes

$$A_i' = \xi'^\dagger\sigma_i\xi'.$$

For example, under a rotation by the angle θ_z around the z-axis the x component transforms as

$$
\begin{aligned}
A_x' &= \xi^\dagger e^{-\frac{i}{2}\sigma_z\theta_z}\sigma_x e^{\frac{i}{2}\sigma_z\theta_z}\xi \\
&= \xi^\dagger\left(\cos\frac{\sigma_z\theta_z}{2} - i\sin\frac{\sigma_z\theta_z}{2}\right)\sigma_x\left(\cos\frac{\sigma_z\theta_z}{2} + i\sin\frac{\sigma_z\theta_z}{2}\right)\xi.
\end{aligned}
$$

Since the expansion of $\cos x$ in a power series contains only even powers of x, and $\sigma_i^2 = 1$, we obtain

$$\cos\frac{\sigma_z\theta_z}{2} = I\cos\frac{\theta_z}{2}.$$

Similarly, it is easy to see that

$$\sin\frac{\sigma_z\theta_z}{2} = \sigma_z\sin\frac{\theta_z}{2},$$

because the expansion of $\sin x$ contains only odd powers of x, and an odd power of σ_z equals σ_z. Hence,

$$
\begin{aligned}
A_x' &= \xi^\dagger\left(I\cos\frac{\theta_z}{2} - i\sigma_z\sin\frac{\theta_z}{2}\right)\left(\sigma_x\cos\frac{\theta_z}{2} + i\sigma_x\sigma_z\sin\frac{\theta_z}{2}\right)\xi \\
&= \xi^\dagger\left(\sigma_x\cos^2\frac{\theta_z}{2} + \sigma_y\sin\frac{\theta_z}{2}\cos\frac{\theta_z}{2} + \sigma_y\sin\frac{\theta_z}{2}\cos\frac{\theta_z}{2} - \sigma_z\sin^2\frac{\theta_z}{2}\right)\xi \\
&= \xi^\dagger(\sigma_x\cos\theta + \sigma_y\sin\theta)\xi = A_x\cos\theta + A_y\sin\theta
\end{aligned}
$$

transforms as the x-component of a three-dimensional vector. Extending this analysis to other components and other rotations we can demonstrate that $A_i = \xi^\dagger \sigma_i \xi$ rotates as a usual vector in three-dimensional space. The only difference is that in the case of space reflections it behaves as a pseudovector, i.e. it does not change its sign.

So far we have considered the representations of the three-dimensional rotation group. It is a three-parameter $SO(3)$ subgroup of the Lorentz group. We shall now construct a representation of the proper Lorentz transformations (boosts along the z-axis).

Suppose that a reference frame moves along the z-axis with velocity v. Then

$$z' = \frac{z + vt}{\sqrt{1 - v^2}} = z \cosh \chi + t \sinh \chi ,$$

$$t' = \frac{t + vz}{\sqrt{1 - v^2}} = z \sinh \chi + t \cosh \chi ,$$

$$(2.11)$$

where $\tanh \chi = v$. These transformations are identical to rotations by a complex angle. Using this correspondence, we choose a two-dimensional representation in the form

$$u_z = e^{\frac{\chi}{2}\sigma_z}.$$

Alternatively, one could choose

$$\tilde{u}_z = e^{-\frac{\chi}{2}\sigma_z}.$$

Then

$$\xi' = e^{\frac{\chi}{2}\sigma_z}\xi , \qquad \xi'^\dagger = \xi^\dagger e^{\frac{\chi}{2}\sigma_z}. \qquad (2.12)$$

It is easy to demonstrate that $A_z = \xi^\dagger \sigma_z \xi$ transforms as the coordinate z in (2.11), while $A_0 = \xi^\dagger \xi$ plays the rôle of the time component of a four-vector, and transforms as the time variable in (2.11). Indeed, from (2.12) we get

$$\begin{aligned} A_0' &= \xi'^\dagger \xi' = \xi^\dagger e^{\chi \sigma_z} \xi = \xi^\dagger (\cosh \chi + \sigma_z \sinh \chi) \xi \\ &= (\xi^\dagger \xi) \cosh \chi + (\xi^\dagger \sigma_z \xi) \sinh \chi = A_0 \cosh \chi + A_z \sinh \chi . \end{aligned}$$

In a similar way we can consider transformation laws for all components (A_0, A_i) under arbitrary boosts and prove that the four-component object $(A_0, \mathbf{A}) = (\xi^\dagger \xi, \xi^\dagger \boldsymbol{\sigma} \xi)$ behaves as a four-vector under Lorentz transformations.

Hence, for the motion along an arbitrary direction $\mathbf{n}$, we may write

$$\xi' = e^{\frac{\chi}{2}(\boldsymbol{\sigma} \cdot \mathbf{n})}\xi. \qquad (2.13)$$

There is another representation of the Lorentz group given by the transformations

$$\dot{\xi}' = e^{-\frac{\chi}{2}(\boldsymbol{\sigma} \cdot \mathbf{n})}\dot{\xi}. \qquad (2.14)$$

(Objects which transform according to (2.14) are marked by dots.) This reflects the fact that in two-dimensional complex space two inequivalent representations of the Lorentz group are realized. The two-component vectors ξ in this space are called spinors. The transformation law (2.14) corresponds to motion in the opposite direction (indeed, the replacement $\chi \to -\chi$ leads to the change of sign of the velocity in (2.11)).

We now have two kinds of spinors, ξ and $\dot{\xi}$, which transform according to different representations of the Lorentz group. Which one should be chosen as the wave function of a spin $\frac{1}{2}$ particle?

Consider a reference frame in which the particle is moving with velocity $\mathbf{v}$

$$\mathbf{v} = \frac{\mathbf{p}}{p_0}, \qquad \frac{1}{\sqrt{1 - \mathbf{v}^2}} = \frac{p_0}{m} = \cosh \chi.$$

It follows from (2.13) that

$$\begin{aligned}
\xi' &= \left[\cosh \frac{\chi}{2} + (\boldsymbol{\sigma} \cdot \mathbf{n}) \sinh \frac{\chi}{2} \right] \varphi e^{-ipx} \\
&= \left[\sqrt{\frac{\cosh \chi + 1}{2}} + (\boldsymbol{\sigma} \cdot \mathbf{n}) \sqrt{\frac{\cosh \chi - 1}{2}} \right] \varphi e^{-ipx}.
\end{aligned}$$

If we accept (2.13) as the law of transformation for the wave function, we get for the moving particle

$$\xi' = \left[\sqrt{\frac{p_0 + m}{2m}} + (\boldsymbol{\sigma} \cdot \mathbf{n}) \sqrt{\frac{p_0 - m}{2m}} \right] \varphi e^{-ipx}. \tag{2.15}$$

There is, however, another possibility. We can choose the transformation law for the wave function as in (2.14). In a sense, it corresponds to a particle moving in the opposite direction. Then

$$\dot{\xi}' = \left[\sqrt{\frac{p_0 + m}{2m}} - (\boldsymbol{\sigma} \cdot \mathbf{n}) \sqrt{\frac{p_0 - m}{2m}} \right] \varphi e^{-ipx}. \tag{2.16}$$

Which of these wave functions should be used to describe spin $\frac{1}{2}$ physical particles has to be decided by experiment (as is the case for massless neutrinos).

Now consider the reflection $\mathbf{n} \to -\mathbf{n}$. Obviously, under this transformation

$$\xi \to \dot{\xi}.$$

Suppose that our electron state has a 'screw'. Such a particle can be described by one of the wave functions ξ, $\dot{\xi}$ in a right-handed coordinate system. The fact that this wave function will change under the transformation to a left-handed coordinate system poses no problem, since the particle itself contains the notion of left and right – which is exactly the case for the neutrino.

If, on the other hand, the particle is completely symmetrical (i.e. it does not know about left and right), then the reflection should not change anything. The description in left- and right-hand reference frames should be equivalent; there is parity conservation. Such particles could be described by a certain superposition of ξ and $\dot{\xi}$. However, one then has to make sure that the difference between ξ and $\dot{\xi}$ does not enter into physical observables.

It proves to be more convenient to introduce instead a four-component wave function and to write it in the form

$$\Psi \propto \begin{pmatrix} \xi \\ \dot{\xi} \end{pmatrix} = \begin{pmatrix} \xi_1 \\ \xi_2 \\ \dot{\xi}_1 \\ \dot{\xi}_2 \end{pmatrix}.$$

If all four components entered all the equations symmetrically, parity would be automatically conserved.

A four-component form of the wave function is not imposed by nature, it is just a convenient way to describe spin $\frac{1}{2}$ particles (which was initiated by the Dirac equation). Also, it is often convenient to introduce two-component functions with definite parities:

$$\begin{aligned} \Psi_1 &= \frac{1}{2}(\xi + \dot{\xi}), \\ \Psi_2 &= \frac{1}{2}(\xi - \dot{\xi}). \end{aligned} \tag{2.17}$$

The explicit form of these functions is

$$\begin{aligned} \Psi_1 &= \sqrt{\frac{p_0 + m}{2m}}\, \varphi\, e^{-ipx}, \\ \Psi_2 &= (\boldsymbol{\sigma} \cdot \mathbf{n})\sqrt{\frac{p_0 - m}{2m}}\, \varphi\, e^{-ipx} = \frac{(\boldsymbol{\sigma} \cdot \mathbf{p})}{p_0 + m}\sqrt{\frac{p_0 + m}{2m}}\, \varphi\, e^{-ipx}. \end{aligned} \tag{2.18}$$

It turns out to be convenient to exclude $\mathbf{n}$:

$$\mathbf{n} = \frac{\mathbf{p}}{|\mathbf{p}|} = \frac{\mathbf{p}}{\sqrt{p_0^2 - m^2}}.$$

Then the four-component wave function becomes

$$\Psi = \begin{pmatrix} \Psi_1 \\ \Psi_2 \end{pmatrix}; \qquad \Psi_2 = \frac{(\boldsymbol{\sigma} \cdot \mathbf{p})}{p_0 + m}\Psi_1. \tag{2.19}$$

Since we have introduced two extra components Ψ_2 merely to preserve the reflection symmetry, they, of course, are not additional degrees of freedom but can be expressed in terms of Ψ_1.

Let us now try to find an equation connecting all four components of the wave function in such a way that there is no difference between right and left and that the particle has a definite (e.g. positive) parity in the rest frame (as follows from (2.19), $\Psi_2 = 0$ at rest). For this purpose, we introduce four-dimensional γ-matrices in the standard representation:

$$\gamma_0 = \begin{pmatrix} I & 0 \\ 0 & -I \end{pmatrix}, \qquad \gamma_i = \begin{pmatrix} 0 & \sigma_i \\ -\sigma_i & 0 \end{pmatrix}.$$

Now it is easy to see that the wave function Ψ (2.19) satisfies the Dirac equation

$$(\gamma_0 p_0 - \boldsymbol{\gamma} \cdot \mathbf{p} - m)\Psi = 0. \tag{2.20}$$

This equation leads to

$$(p_0 - m)\Psi_1 - (\boldsymbol{\sigma} \cdot \mathbf{p})\Psi_2 = 0, \tag{2.21}$$
$$(-p_0 - m)\Psi_2 + (\boldsymbol{\sigma} \cdot \mathbf{p})\Psi_1 = 0. \tag{2.22}$$

Obviously, the relation between Ψ_2 and Ψ_1 given in (2.19) follows from (2.22). Substituting Ψ_2 into (2.21) we obtain the standard relativistic relation between energy and momentum:

$$\left[(p_0 - m) - \frac{(\boldsymbol{\sigma} \cdot \mathbf{p})^2}{p_0 + m} \right] \Psi_1 = 0$$

or

$$\frac{p_0^2 - m^2 - \mathbf{p}^2}{p_0 + m} \Psi_1 = 0,$$

i.e.

$$p_0^2 - \mathbf{p}^2 = m^2.$$

Equation (2.20) is relativistically invariant because so is the scalar product of the two four-vectors γ_μ and p_μ:

$$\hat{p} \equiv \gamma_\mu p_\mu = \gamma_0 p_0 - \boldsymbol{\gamma} \cdot \mathbf{p}.$$

Thus

$$(\hat{p} - m)\Psi = 0 \tag{2.23}$$

selects and describes the states with positive internal parity in the particle rest frame, because only Ψ_1 does not vanish at $\mathbf{v} = 0$.

We could have written instead

$$(\hat{p} + m)\Psi = 0, \tag{2.24}$$

which equation, unlike the standard Dirac equation (2.23), would select *negative* parity states of the particle at rest.

To extend our description to the case of massless spin $\frac{1}{2}$ particles we have to get rid of the masses in the denominators of the wave functions (2.18). This can be done by changing the normalization of the Dirac wave function in (2.20). We multiply it by $\sqrt{2m}$ and write the new wave function in terms of the spinors u^λ

$$\Psi^\lambda = \begin{pmatrix} \sqrt{p_0 + m}\,\varphi_\lambda \\ (\boldsymbol{\sigma}\cdot\mathbf{n})\sqrt{p_0-m}\,\varphi_\lambda \end{pmatrix} e^{-ipx} = \sqrt{p_0+m}\begin{pmatrix} \varphi_\lambda \\ \frac{(\boldsymbol{\sigma}\cdot\mathbf{p})}{p_0+m}\varphi_\lambda \end{pmatrix} e^{-ipx} = u^\lambda(p)e^{-ipx},$$

(2.25)

where $\bar{u}^\lambda u^\lambda = 2m$. Since φ has two components, there are two linearly independent φ_λ, ($\lambda = \pm 1$), corresponding to two spin projections. In the rest frame $\frac{\lambda}{2}$ is simply the projection of spin

$$\sigma_z \varphi_\lambda = \lambda \varphi_\lambda.$$

(2.26)

It is easy to see that

$$\varphi_{+1} = \begin{pmatrix} 1 \\ 0 \end{pmatrix}, \qquad \varphi_{-1} = \begin{pmatrix} 0 \\ 1 \end{pmatrix},$$

and the wave function $\begin{pmatrix} a_1 \\ a_2 \end{pmatrix}$ in the rest frame can be written in the form

$$\begin{pmatrix} a_1 \\ a_2 \end{pmatrix} = a_1 \begin{pmatrix} 1 \\ 0 \end{pmatrix} + a_2 \begin{pmatrix} 0 \\ 1 \end{pmatrix}.$$

The functions a_1 and a_2 are the probability amplitudes for a particle to have spin projections $+\frac{1}{2}$ and $-\frac{1}{2}$.

Let us write (2.26) in a relativistically covariant form. First, we introduce in the rest frame a unit vector $\boldsymbol{\zeta}$ directed along the spin. Then (2.26) may be written as

$$(\boldsymbol{\sigma} \cdot \boldsymbol{\zeta})\varphi = \lambda\varphi.$$

(2.27)

Introducing a space-like four-vector ζ_μ ($\zeta_\mu^2 = -1$) which in the rest frame turns into $(0, \boldsymbol{\zeta})$, and the four-matrix

$$\gamma_5 = \begin{pmatrix} 0 & I \\ I & 0 \end{pmatrix},$$

we can write a relativistically invariant expression which corresponds to (2.27):

$$(\gamma_5 \zeta_\mu \gamma_\mu - \lambda)u = 0.$$

(2.28)

Indeed, in the rest frame

$$-(\gamma_5\zeta_i\gamma_i + \lambda)u = [(\boldsymbol{\sigma}\cdot\boldsymbol{\zeta}) - \lambda]\varphi = 0.$$

Here we used that the lower components of u are zero in the rest frame and

$$\gamma_5\gamma_i = \begin{pmatrix} -\sigma_i & 0 \\ 0 & \sigma_i \end{pmatrix}.$$

Then (2.28) turns into (2.27). Hence, in order to define the Dirac spinor unambiguously, two equations are necessary:

$$\begin{aligned} (\hat{p} - m)u &= 0, \\ (\gamma_5\hat{\zeta} - \lambda)u &= 0. \end{aligned} \tag{2.29}$$

We shall denote the solution of these equations by either $u(p, \zeta)$ or $u^\lambda(p)$ (where λ and ζ are fixed).

Let us now establish a probabilistic interpretation for the spinor wave functions. As usual, we need to construct a conserved quantity. Obviously, the product of spinors of different types which transform according to different representations of the Lorentz group is relativistically invariant. Indeed,

$$\dot{\xi}'^\dagger\xi' = \dot{\xi}^\dagger e^{-\frac{\chi}{2}(\boldsymbol{\sigma}\cdot\mathbf{n})}e^{\frac{\chi}{2}(\boldsymbol{\sigma}\cdot\mathbf{n})}\xi = \dot{\xi}^\dagger\xi.$$

However, we use not the spinors ξ and $\dot{\xi}$ but their linear combinations

$$\begin{aligned} \Psi_1 &= \frac{1}{2}(\xi + \dot{\xi}), \\ \Psi_2 &= \frac{1}{2}(\xi - \dot{\xi}). \end{aligned}$$

In these terms the relativistically invariant product has the form

$$(\dot{\xi}^\dagger\xi) = (\Psi_1^\dagger - \Psi_2^\dagger, \Psi_1 + \Psi_2) = \Psi_1^\dagger\Psi_1 - \Psi_2^\dagger\Psi_2.$$

Introduce the Dirac conjugate four-component spinor

$$\bar{u}^\lambda(p) \equiv u^\dagger(p)\gamma_0 = (\Psi_1^\dagger, \Psi_2^\dagger)\begin{pmatrix} I & 0 \\ 0 & -I \end{pmatrix} = (\Psi_1^\dagger, -\Psi_2^\dagger).$$

Then the product

$$\bar{u}^\lambda(p)u^{\lambda'}(p) = \Psi_1^\dagger\Psi_1' - \Psi_2^\dagger\Psi_2' \tag{2.30}$$

is a relativistic invariant.

Let us find an equation for the Dirac conjugate spinor $\bar{u}^\lambda$. The first of equations (2.29) gives

$$u^\dagger(\hat{p}^\dagger - m) = 0, \tag{2.31}$$

where
$$\hat{p}^\dagger = (\gamma_0 p_0 - \boldsymbol{\gamma} \cdot \mathbf{p})^\dagger = \gamma_0 p_0 + \boldsymbol{\gamma} \cdot \mathbf{p}$$

due to hermiticity of γ_0 and the antihermiticity of γ_i. Multiplying (2.31) by γ_0 from the right and using the commutation law for the γ-matrices

$$\gamma_\mu \gamma_\nu + \gamma_\nu \gamma_\mu = 2 g_{\mu\nu},$$

i.e.

$$\gamma \gamma_0 = -\gamma_0 \gamma,$$

we obtain

$$\bar{u}(\hat{p} - m) = 0. \tag{2.32}$$

And what about the second of equations (2.29)? One has, as in the previous case,

$$u^\dagger(\hat{\zeta}^\dagger \gamma_5 - \lambda) = 0$$

which gives, after multiplication from the right by γ_0,

$$\bar{u}(\gamma_5 \hat{\zeta} - \lambda) = 0. \tag{2.33}$$

In other words, $\bar{u}^\lambda$ and u^λ are solutions of the same equations.

As we have seen, $\bar{u}u$ is a relativistic invariant, and γ_μ transforms like a four-vector. Then $j_\mu = \bar{u}\gamma_\mu u$ also transforms like a four-vector. Its zeroth component $\bar{u}\gamma_0 u = u^+ u$ can be identified with the probability density, and $\bar{u}\gamma_i u$ with the probability current density. Indeed, in the coordinate representation equations (2.29) and (2.32) can be written as

$$\left(i\frac{\overrightarrow{\partial}}{\partial x_\mu}\gamma_\mu - m \right) \Psi(x) \;=\; 0, \tag{2.34}$$

$$\bar{\Psi}(x) \left(-i\frac{\overleftarrow{\partial}}{\partial x_\mu}\gamma_\mu - m \right) \;=\; 0. \tag{2.35}$$

(The arrows here denote the direction of the differentiation.) To confirm that j_μ obeys the equation of continuity and j_0 can be considered as the probability density, we multiply (2.34) by $\bar{\Psi}$ from the left and (2.35) by Ψ from the right. Subtracting these equations, we obtain the local conservation law

$$\frac{\partial}{\partial x_\mu}(\bar{\Psi}\gamma_\mu\Psi) = 0. \tag{2.36}$$

As usual, particles correspond to the positive-frequency solutions, and we have to learn how to construct such solutions of the Dirac equation. The positive- and negative-frequency solutions differ by the substitution

$$p_0 \to -p_0, \qquad \mathbf{p} \to -\mathbf{p}. \tag{2.37}$$

Our spinor has the form

$$u^\lambda(p) = \begin{pmatrix} \sqrt{p_0 + m}\varphi_\lambda \\ (\boldsymbol{\sigma} \cdot \mathbf{n})\sqrt{p_0 - m}\varphi_\lambda \end{pmatrix}. \qquad (2.38)$$

Performing the replacement (2.37), we get

$$u^\lambda(-p) = \pm i \begin{pmatrix} \sqrt{p_0 - m}\varphi_\lambda \\ (\boldsymbol{\sigma} \cdot \mathbf{n})\sqrt{p_0 + m}\varphi_\lambda \end{pmatrix}. \qquad (2.39)$$

What is the connection between $u^\lambda(p)$ and $u^\lambda(-p)$? For scalar particles it was trivial: $\varphi(-p) = \varphi^*(p)$. Consider

$$
\begin{aligned}
u^{\lambda*}(-p) &= \mp i \begin{pmatrix} \sqrt{p_0 - m}\varphi_\lambda^* \\ (\boldsymbol{\sigma} \cdot \mathbf{n})^*\sqrt{p_0 + m}\varphi_\lambda^* \end{pmatrix} \\
&\equiv \mp i \begin{pmatrix} \sigma_y\sqrt{p_0 - m}\sigma_y\varphi_\lambda^* \\ -\sigma_y(\boldsymbol{\sigma} \cdot \mathbf{n})\sqrt{p_0 + m}\sigma_y\varphi_\lambda^* \end{pmatrix},
\end{aligned} \qquad (2.40)
$$

where we have multiplied both components by $\sigma_y^2 = 1$ and used the relations $\sigma_y^* = -\sigma_y$ and $\sigma_i^* = \sigma_i$, $\sigma_i\sigma_y = -\sigma_y\sigma_i$ for $i = x, z$.

Let us define a new spinor $\varphi_\lambda' \equiv (\boldsymbol{\sigma} \cdot \mathbf{n})\sigma_y\varphi_\lambda^*$. Then, because of $(\boldsymbol{\sigma} \cdot \mathbf{n})(\boldsymbol{\sigma} \cdot \mathbf{n}) = \mathbf{n}^2 = 1$, we have $\sigma_y\varphi_\lambda^* = (\boldsymbol{\sigma} \cdot \mathbf{n})\varphi_\lambda'$, and

$$
\begin{aligned}
u^{\lambda*}(-p) &= \pm \begin{pmatrix} -i\sigma_y(\boldsymbol{\sigma} \cdot \mathbf{n})\sqrt{p_0 - m}\,\varphi_\lambda' \\ i\sigma_y\sqrt{p_0 + m}\,\varphi_\lambda' \end{pmatrix} \\
&= \pm \begin{pmatrix} 0 & -i\sigma_y \\ i\sigma_y & 0 \end{pmatrix} \begin{pmatrix} \sqrt{p_0 + m}\,\varphi_\lambda' \\ (\boldsymbol{\sigma} \cdot \mathbf{n})\sqrt{p_0 - m}\,\varphi_\lambda' \end{pmatrix}.
\end{aligned} \qquad (2.41)
$$

The last column resembles the original four-spinor (2.38). What is φ'? Let us show that φ_λ' describes a particle in a state with the opposite spin. This means that if φ_λ satisfies the equation

$$(\boldsymbol{\sigma} \cdot \mathbf{n})\,\varphi_\lambda = \lambda\varphi_\lambda, \qquad (2.42)$$

then for φ_λ' we have

$$(\boldsymbol{\sigma} \cdot \mathbf{n})\,\varphi_\lambda' = -\lambda\,\varphi_\lambda'. \qquad (2.43)$$

It follows from (2.42) that

$$(\boldsymbol{\sigma}^* \cdot \mathbf{n})\,\varphi_\lambda^* = \lambda\varphi_\lambda^*.$$

Multiplying this expression by σ_y from the left and using

$$\sigma_y\varphi_\lambda^* = (\boldsymbol{\sigma} \cdot \mathbf{n})\,\varphi_\lambda',$$

we get

$$\sigma_y(\boldsymbol{\sigma}^* \cdot \mathbf{n})\varphi_\lambda^* = \lambda(\boldsymbol{\sigma} \cdot \mathbf{n})\,\varphi_\lambda'.$$

On the other hand,

$$\sigma_y(\boldsymbol{\sigma}^* \cdot \mathbf{n})\varphi_\lambda^* = -(\boldsymbol{\sigma} \cdot \mathbf{n})\sigma_y\varphi_\lambda^* = -\varphi_\lambda'.$$

This means that

$$-\varphi_\lambda' = \lambda(\boldsymbol{\sigma} \cdot \mathbf{n})\,\varphi_\lambda'.$$

Since $\lambda = \pm 1$ we have

$$(\boldsymbol{\sigma} \cdot \mathbf{n})\,\varphi_\lambda' = -\lambda\,\varphi_\lambda',$$

i.e. indeed, $\varphi_\lambda' = \varphi_{-\lambda}$.

Using this fact we can now construct the Dirac conjugated spinor to $u^\lambda(-p)$. Transposing $u^{\lambda*}(-p)$ given in (2.41) and multiplying by γ_0 results in

$$\overline{u^\lambda(-p)} = \pm[u^{-\lambda}(p)]^\top \begin{pmatrix} 0 & -i\sigma_y \\ i\sigma_y & 0 \end{pmatrix} \begin{pmatrix} I & 0 \\ 0 & -I \end{pmatrix} = [u^{-\lambda}(p)]^\top C, \quad (2.44)$$

where

$$C = -i\gamma_2\gamma_0 = \begin{pmatrix} 0 & i\sigma_y \\ i\sigma_y & 0 \end{pmatrix} \qquad (2.45)$$

is called the charge conjugation matrix. It has the following properties:

$$C^2 = -1, \quad C^\dagger = C^\top = C^{-1} = -C,$$

$$C\gamma^\mu C^{-1} = -\gamma_\mu^\top. \qquad (2.46)$$

(Note that according to (2.44) the matrix C is defined up to a sign.) Thus the connection between positive- and negative-frequency Dirac spinors is given by

$$\overline{u^\lambda(-p)} = [u^{-\lambda}(p)]^\top C. \qquad (2.47)$$

It is convenient to introduce four-spinor v^λ:

$$v^\lambda(p) = u^\lambda(-p).$$

Let us find the connection between v and $\bar{u}$. Multiplying (2.47) by γ_0,

$$u^{\lambda\dagger}(-p) = [u^{-\lambda}(p)]^\top C\gamma_0,$$

and taking the Hermitian conjugate we get

$$u^\lambda(-p) = \gamma_0^\dagger C^\dagger[u^{-\lambda\dagger}(p)]^\top = \gamma_0 C^\dagger \gamma_0[u^{-\lambda\dagger}(p)\gamma_0]^\top = C[\bar{u}^{-\lambda}(p)]^\top,$$

i.e.

$$v^\lambda(p) = C[\bar{u}^{-\lambda}(p)]^\top = -[\bar{u}^{-\lambda}(p)\,C]^\top,$$

$$\bar{v}^\lambda(p) = [u^{-\lambda}(p)]^\top C = [C^{-1}u^{-\lambda}(p)]^\top. \qquad (2.48)$$

We have obtained $u^\lambda(-p)$ in (2.39) as a result of the substitution $p \to -p$ in the four-spinor $u^\lambda(p)$ in (2.38). What is the connection between the respective Dirac conjugated spinors? Does the substitution $p \to -p$ in $\bar{u}^\lambda(p)$ lead to the spinor $u^\lambda(-p)$ in (2.44)? From (2.25) we have

$$\bar{u}^\lambda(p) = \sqrt{p_0 + m}\left(\varphi_\lambda^\dagger, \; -\varphi_\lambda^\dagger \frac{(\boldsymbol{\sigma}\cdot\mathbf{p})^\dagger}{p_0 + m}\right) = \left(\sqrt{p_0+m}\,\varphi_\lambda^\dagger, \; -\varphi_\lambda^\dagger(\boldsymbol{\sigma}\cdot\mathbf{n})\sqrt{p_0-m}\right).$$

The substitution results in

$$\bar{u}^\lambda(-p) = \pm i\left(\sqrt{p_0 - m}\,\varphi_\lambda^\dagger, \; -\varphi_\lambda^\dagger(\boldsymbol{\sigma}\cdot\mathbf{n})\sqrt{p_0 + m}\right).$$

On the other hand, (2.40) gives

$$\begin{aligned}
\overline{u^\lambda(-p)} &\equiv \left[u^\lambda(-p)\right]^\top \gamma_0 = \mp i\begin{pmatrix} \sqrt{p_0 - m}\,\varphi_\lambda^* \\ (\boldsymbol{\sigma}\cdot\mathbf{n})^*\sqrt{p_0 + m}\,\varphi_\lambda^* \end{pmatrix}^\top \gamma_0 \\
&= \mp i\left(\sqrt{p_0 - m}\,\varphi_\lambda^\dagger, \; -\varphi_\lambda^\dagger(\boldsymbol{\sigma}\cdot\mathbf{n})\sqrt{p_0 + m}\right).
\end{aligned}$$

Comparing these expressions we see that

$$\bar{u}^\lambda(-p) = -\overline{u^\lambda(-p)}.$$

Here $\bar{u}^\lambda(-p)$ is the function $\bar{u}^\lambda(p)$ after the substitution $p \to -p$, while $\overline{u^\lambda(-p)}$ is the Dirac conjugate of the function $u^\lambda(-p)$. Thus, the functions $\bar{u}^\lambda(-p)$ and $\overline{u^\lambda(-p)}$ do not coincide but differ by sign. Therefore, since

$$v^\lambda(p) = u^\lambda(-p),$$

we have

$$\bar{v}^\lambda(p) = -\bar{u}^\lambda(-p). \tag{2.49}$$

This means that the solutions $u^\lambda(p)$ and $\bar{u}^\lambda(p)$ cease to be Dirac conjugated after changing the sign of momentum, $p \to -p$.

In what follows we will need two useful relations:

(1) The normalization condition

$$\bar{u}_\alpha^\lambda(p)u_\alpha^{\lambda'}(p) = 2m\delta_{\lambda\lambda'}. \tag{2.50}$$

(Here α enumerates the four components of the Dirac spinor.)

This equality follows directly from the explicit form of the four-component spinors

$$\begin{aligned}
\bar{u}_\alpha^\lambda(p)u_\alpha^{\lambda'}(p) &= \left(\sqrt{p_0 + m}\,\varphi_\lambda^+, \; -\varphi_\lambda^+(\boldsymbol{\sigma}\cdot\mathbf{n})\sqrt{p_0 - m}\right)\begin{pmatrix} \sqrt{p_0 + m}\,\varphi_{\lambda'} \\ (\boldsymbol{\sigma}\cdot\mathbf{n})\sqrt{p_0 - m}\,\varphi_{\lambda'} \end{pmatrix} \\
&= (p_0 + m - p_0 + m)(\varphi_\lambda^+ \varphi_{\lambda'}) = 2m\delta_{\lambda\lambda'}.
\end{aligned}$$

(2) The completeness relation

$$\sum_{\lambda=1,2} u_\alpha^\lambda(p)\bar{u}_\beta^\lambda(p) = (\hat{p}+m)_{\alpha\beta}. \qquad (2.51)$$

This can be proved with the help of the identity

$$\sum_{\lambda=1,2} u_\alpha^\lambda(p)\bar{u}_\beta^\lambda(p) = \frac{1}{2m}\sum_{\lambda=1}^{4}(\hat{p}+m)_{\alpha\gamma}u_\gamma^\lambda\bar{u}_\beta^\lambda,$$

where we have introduced a summation over two additional states with negative parities. This identity is valid because u^λ are solutions of the Dirac equation and satisfy the completeness relation

$$\sum_{\lambda=1}^{4} u_\gamma^\lambda\bar{u}_\beta^\lambda = 2m\delta_{\gamma\beta}.$$

2.2 The Green function of the electron

For a spin $\frac{1}{2}$ particle we obtained the Dirac equation

$$(\gamma_\mu p_\mu - m)\Psi(p) = 0 \qquad (2.52)$$

or, in the coordinate representation,

$$\left(i\gamma_\mu\frac{\partial}{\partial x_\mu} - m\right)\Psi(x) = 0. \qquad (2.53)$$

The Green function $G(x)$ satisfies the equation

$$\left(i\gamma_\mu\frac{\partial}{\partial x_\mu} - m\right)G(x) = i\delta(x). \qquad (2.54)$$

In the momentum space we get, as usual,

$$(\hat{p} - m)G(p) = -1$$

and

$$G(p) = \frac{1}{m - \hat{p} - i\epsilon} = \frac{m + \hat{p}}{m^2 - p^2 - i\epsilon} \qquad (2.55)$$

where the relation

$$\hat{p}\hat{p} = \gamma_\mu p_\mu \gamma_\nu p_\nu = \frac{1}{2}(\gamma_\nu\gamma_\mu + \gamma_\nu\gamma_\mu)p_\mu p_\nu = p^2$$

was used. Then

$$G_{\alpha\beta}(x_2 - x_1) = \int \frac{d^4p}{(2\pi)^4 i} \frac{(m + \hat{p})_{\alpha\beta}}{m^2 - p^2 - i\varepsilon} e^{-ip(x_2-x_1)}. \qquad (2.56)$$

Let us calculate $G(x_2 - x_1)$ when $t_2 > t_1$. Will we obtain from this integral

$$\begin{array}{ccc}
\alpha & & \beta \\
\hline
x_1 & \longrightarrow & x_2
\end{array}$$

i.e. the electron propagator from x_1 to x_2? Taking the residue at the pole $p_0 = \sqrt{m^2 + \mathbf{p}^2}$ we have

$$\begin{aligned}
G_{\beta\alpha}(x_2 - x_1) &= \int \frac{d^3p}{(2\pi)^3 2p_0} e^{-ip(x_2-x_1)}(m + \hat{p})_{\beta\alpha} \\
&= \int \frac{d^3p}{(2\pi)^3 2p_0} e^{-ip(x_2-x_1)} \sum_{\lambda} u_{\beta}^{\lambda}(p) \bar{u}_{\alpha}^{\lambda}(p).
\end{aligned}$$

Introducing the electron wave function

$$\Psi_{\alpha}^{\lambda}(p, x) = \frac{u_{\alpha}^{\lambda}(p)}{\sqrt{2p_0}} e^{-ipx}, \qquad (2.57)$$

we have

$$G_{\beta\alpha}(x_2 - x_1) = \sum_{\lambda=1,2} \int \frac{d^3p}{(2\pi)^3} \Psi_{\beta}^{\lambda}(p, x_2) \bar{\Psi}_{\alpha}^{\lambda}(p, x_1) \qquad (2.58)$$

which is indeed the electron propagator describing propagation of positive frequencies.

What will happen if $t_2 < t_1$? In this case the contour has to be closed around the other pole, $p_0 = -\sqrt{m^2 + \mathbf{p}^2}$, and we obtain

$$\begin{aligned}
G_{\beta\alpha}(x_2 - x_1) &= \int \frac{d^3p}{(2\pi)^3 2p_0} e^{ip(x_2-x_1)}(m - \hat{p})_{\beta\alpha} \\
&= -\sum_{\lambda} \int \frac{d^3p}{(2\pi)^3 2p_0} v_{\beta}^{\lambda}(p) \bar{v}_{\alpha}^{\lambda}(p) e^{ip(x_2-x_1)} \\
&= -\sum_{\lambda} \int \frac{d^3p}{(2\pi)^3} \Psi_{\beta}^{\lambda-}(p, x_2) \bar{\Psi}_{\alpha}^{\lambda-}(p, x_1), \qquad (2.59)
\end{aligned}$$

where we have changed the sign of the integration three-momentum $\mathbf{p}$ and defined the positive-energy four-momentum $p = (p_0, \mathbf{p})$, $p_0 = \sqrt{m^2 + \mathbf{p}^2}$, as in (2.58). We also introduced

$$\Psi_{\alpha}^{\lambda-}(p, x) = \frac{v_{\alpha}^{\lambda}(p)}{\sqrt{2p_0}} e^{ipx} \qquad (2.60)$$

and used the completeness relation (2.51),

$$\sum_\lambda u_\alpha^\lambda(-p)\bar{u}_\beta^\lambda(-p) = (m - \hat{p})_{\alpha\beta},$$

and

$$u_\alpha^\lambda(-p) = v_\alpha^\lambda(p), \qquad \bar{u}_\beta^\lambda(-p) = -\bar{v}_\beta^\lambda(p).$$

It follows from (2.59) that a negative frequency state propagates backwards in time (from x_1 to x_2). It can also be interpreted as the propagation of an antiparticle (positron in our case), described by the function $\Psi^-(p, x)$, forward in time i.e. from x_2 to x_1.

Recall now that the charge conjugation matrix C has the property

$$C\gamma_\mu C^{-1} = -\gamma_\mu^\mathsf{T},$$

and hence,

$$(m - \hat{p})_{\beta\alpha} = [C(m + \hat{p})C^{-1}]_{\alpha\beta}.$$

Then we easily obtain an identity for the Green functions

$$G_{\beta\alpha}(x_2 - x_1) = \int \frac{d^4p}{(2\pi)^4 i} e^{-ip(x_1 - x_2)} \frac{[C(m + \hat{p})C^{-1}]_{\alpha\beta}}{m^2 - p^2},$$

or

$$G^\mathsf{T}(x_2 - x_1) = C\,G(x_1 - x_2)\,C^{-1}. \tag{2.61}$$

We see that unlike the case of scalar and vector particles (cf. (1.81) and (1.89)), the electron Green function is not symmetric under the transposition of the coordinates, $x_2 - x_1 \to x_1 - x_2$. This complication can be understood by bearing in mind that G is a matrix which undergoes unitary transformation when $x \to -x$. Consequently, the same process is described in another representation.

The Green functions of electrons and positrons turn out to be different and the connection between them is realized by the charge conjugation matrix. The fact that the Green functions of e$^-$ and e$^+$ are different causes no problems since so far we have not seen any spin $\frac{1}{2}$ particles identical to their antiparticles. If such a Majorana-type particle existed, its propagation could be described in the same way as that of a charged particle, but its interaction would not change under charge conjugation. Also, a formalism could be constructed in which the asymmetry in the description of propagation would not arise at all.

2.3 Matrix elements of electron scattering amplitudes

Let us consider the electron–electron scattering process. To calculate its amplitude, it is necessary to take the limit $x_{10}, x_{20} \to -\infty$, $x_{30}, x_{40} \to$

$+\infty$. This time ordering determines unambiguously how to close the contours around the poles in the Green functions corresponding to the external lines. After taking the residues, the external lines correspond to linear combinations of the type $\Psi^\lambda_\alpha(x_3)\bar{\Psi}^\lambda_\beta(x'_3)$ which, unlike the case of scalar particles, are matrices and not numbers.

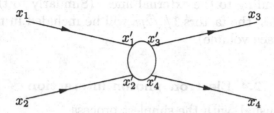

Let us now look at the diagram time-ordered from top to bottom, i.e. take the limit

$$x_{10}, x_{30} \to -\infty, \qquad x_{20}, x_{40} \to +\infty.$$

The Green functions that describe the propagation from x_1 to x'_1 and from x'_4 to x_4 do not change, while the two other Green functions give, according to the change in the direction of closing the loops around the poles, $-\bar{\Psi}^-(x'_3)\Psi^-(x_3)$, and a similar expression for the line $x'_2 - x_2$.

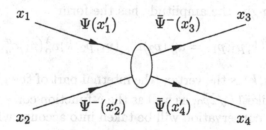

We thus come to the conclusion that (similarly to the case of spin zero particles) the external lines in the scattering amplitudes for particles with non-zero spins correspond to wave functions instead of Green functions. Repeating the calculations (1.130)–(1.131), we see that the transition amplitude differs from the amplitude in the scalar case (1.131) only by the spinor factors:

	u	corresponds to	the initial electron,
	$\bar{u}$	corresponds to	the final electron,
	$\bar{v}$	corresponds to	the initial positron,
and	v	corresponds to	the final positron.

According to (2.59), a factor -1 per positron Green function should be included in the transition amplitude. In the above example we had two antiparticles, so this did not matter. At the same time, the transition amplitude for the pair creation process $\gamma\gamma \to e^- e^+$ where there is but one antiparticle line will have an additional minus sign.

Thus, the transition amplitude $S(p_1, p_2; k_1, k_2)$ for a spin $\frac{1}{2}$ particle can be obtained in the following way: the internal part of the diagram has to be calculated in the momentum representation and multiplied by the spinors corresponding to the external lines. (Similarly to the case of the spin zero particles, the factors $1/\sqrt{2p_0}$ will be included in the expression for the phase space volume).

2.4 Electron–photon interaction

Let us start, as usual, with the simplest process

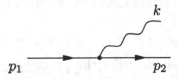

and write the amplitude $T(k, p_2; p_1)$ explicitly. As we have shown, the spinor factors $u^\lambda(p_1)$ and $\bar{u}^{\lambda'}(p_2)$ correspond to the initial- and final-state electrons, respectively. The photon is described by the polarization vector e^σ_μ. Correspondingly, the amplitude has the form

$$T(k, p_2; p_1) = \bar{u}^{\lambda'}_\beta(p_2)\Gamma^\mu_{\beta\alpha}(p_1, p_2, k)u^\lambda_\alpha(p_1)e^\sigma_\mu, \qquad (2.62)$$

where $\Gamma^\mu_{\beta\alpha}(p_1, p_2, k)$ is the vertex (the internal part of the graph). The additional factors like $1/\sqrt{2p_0}$, as well as the δ-function corresponding to the four-momentum conservation will be taken into account when calculating the cross sections.

Let us now construct the vertex function $\Gamma^\mu_{\beta\alpha}$. The amplitude $\bar{u}\Gamma^\mu u$ should be a vector. We have three Lorentz vectors at our disposal, the matrix γ_μ and the two independent momentum vectors, $k = p_1 - p_2$ and $p \equiv p_1 + p_2$, and can write

$$\Gamma^\mu = a\gamma_\mu + bp_\mu + ck_\mu. \qquad (2.63)$$

We could have also tried more complicated structures like $d_1 \cdot \gamma_\mu \hat{p}_1 + d_2 \cdot \hat{p}_2\gamma_\mu$. We have to remember, however, that Γ is sandwiched between two spinors which satisfy the Dirac equation, so that these two structures redefine the parameter a of the γ_μ term in (2.63).

Similar consideration applies to terms with the opposite order of matrices: $d'_1 \cdot \hat{p}_1\gamma_\mu + d'_2 \cdot \gamma_\mu\hat{p}_2$. Here we use the commutation relation

$$\hat{p}_1\gamma_\mu \equiv \gamma_\nu\gamma_\mu p_{1\nu} = -\gamma_\mu\gamma_\nu p_{1\nu} + 2g_{\mu\nu}p_{1\nu} = -\gamma_\mu\hat{p}_1 + 2p_{1\mu} \qquad (2.64)$$

and observe that these two structures are not independent either. They reduce, on the mass shell, to those already present in (2.63) which, therefore, proves to be the most general form of the vertex function.

Let us write (2.63) in a slightly different form. Due to the commutation relation

$$\gamma_\mu \hat{p} + \hat{p}\gamma_\mu = \gamma_\mu \hat{p} - \gamma_\mu \hat{p} + 2p_\mu$$

we have

$$\bar{u}(p_2)[\gamma_\mu \hat{p} + \hat{p}\gamma_\mu]u(p_1) = 2p_\mu \bar{u}(p_2)u(p_1). \tag{2.65}$$

On the other hand,

$$p_1 + p_2 = 2p_1 - k = 2p_2 + k,$$

and

$$\bar{u}(p_2)[\gamma_\mu \hat{p} + \hat{p}\gamma_\mu]u(p_1) = 4m\bar{u}(p_2)\gamma_\mu u(p_1) + \bar{u}(p_2)(\hat{k}\gamma_\mu - \gamma_\mu \hat{k})u(p_1). \tag{2.66}$$

Thus, we can use $\hat{k}\gamma_\mu - \gamma_\mu \hat{k}$ instead of p_μ:

$$\Gamma^\mu = a\gamma_\mu + b(\hat{k}\gamma_\mu - \gamma_\mu \hat{k}) + ck_\mu. \tag{2.67}$$

Let us now determine a, b and c (naturally these factors are different here from those in (2.63)). The photon emission amplitude should satisfy the transversality condition

$$k_\mu (\bar{u}\Gamma^\mu u) = 0. \tag{2.68}$$

It follows from (2.67) that

$$k_\mu \Gamma^\mu = a\hat{k} + ck^2, \tag{2.69}$$

and thus

$$\bar{u}(p_2)k_\mu \Gamma_\mu u(p_1) = a[\bar{u}(p_2)(\hat{p}_1 - \hat{p}_2)u(p_1)] + ck^2\bar{u}(p_2)u(p_1) = 0. \tag{2.70}$$

Since

$$\bar{u}(p_2)(\hat{p}_1 - \hat{p}_2)u(p_1) = \bar{u}(p_2)(m - m)u(p_1) = 0,$$

(2.70) leads to the condition $c = 0$.

Generally speaking, there is no restriction on the constant b. So far there are two experimentally known particles: the electron and the muon, for which $b = 0$ with very high accuracy (although a small effective b is generated *dynamically* even for these particles when one considers more complicated radiation processes).

Usually it is assumed that for the *fundamental* interaction between the elementary fermions and photons

$$b = 0, \quad a = \text{const.} \tag{2.71}$$

This is the hypothesis of a minimal electromagnetic interaction. It is justified, on the one hand, by its simplicity. On the other hand, no self-consistent theory can be constructed if $b \neq 0$. We will show later that $b \neq 0$ corresponds to the description of particles with anomalous magnetic moments.

Hence,

$$\Gamma_\mu = e\,\gamma_\mu. \tag{2.72}$$

(We will see below in Section 2.5.2 that the numerical factor in (2.72) is equal to the charge of the particle.) The invariant amplitude (2.62) for photon emission by an electron can now be written as

$$T_{e^-} = e\,\bar{u}^{\lambda'}(p_2)\hat{e}^\sigma u^\lambda(p_1), \tag{2.73}$$

where

$$\hat{e}^\sigma \equiv \gamma_\mu e^\sigma_\mu.$$

According to our rules for the amplitudes involving antiparticles, photon emission by a positron is described by a similar expression,

$$T_{e^+} = e\,\bar{v}^\lambda(p_1^+)\hat{e}^\sigma v^{\lambda'}(p_2^+), \tag{2.74}$$

where $\bar{v}^\lambda$ corresponds to the initial, and $v^{\lambda'}$ to the final positron with physical (positive-energy) momenta p_1^+ and p_2^+.

Consider how the amplitudes of photon emission by an electron and a positron are related. Recall that (see (2.48))

$$\begin{aligned}
v^\lambda(p) &\equiv u^\lambda(-p) = -[\bar{u}^{-\lambda}(p)C]^\top = C[\bar{u}^{-\lambda}(p)]^\top, \\
\bar{v}^\lambda(p) &\equiv \overline{u^\lambda(-p)} = -\bar{u}^\lambda(-p) = [u^{-\lambda}(p)]^\top C.
\end{aligned} \tag{2.75}$$

This leads to

$$T_{e^+} = e\bar{u}^{-\lambda'}(p_2^+)\hat{e}^\sigma u^{-\lambda}(p_1^+), \tag{2.76}$$

since $C\hat{e}C = [\hat{e}]^\top$ (see (2.46)). Thus, we see that for given initial and final particle momenta, T_{e^-} (2.73) and T_{e^+} (2.76) are equal up to the values of the spin variables (the signs of λ, λ').

Let us now look at the relationship between these two amplitudes from another angle. For this purpose we redraw the diagram which corresponds to photon emission by an electron as

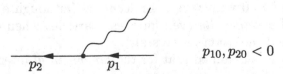

and replace

$$p_1 = -p_2^+, \qquad p_2 = -p_1^+.$$

In this way we obtain the *analytically continued* amplitude T_{cont}

$$T_{\text{cont}} = e\bar{u}(-p_1^+)\hat{e}^\sigma u(-p_2^+) = -e\bar{v}(p_1^+)\hat{e}^\sigma v(p_2^+) \qquad (2.77)$$

which, except for the sign, coincides with the amplitude of photon emission by a positron. This is quite natural, since the change of sign of x in coordinate space is equivalent to the change of sign of p in momentum space, and, of course, we obtain the amplitude of an antiparticle by replacing p by $-p$.

But where has this extra 'minus' come from? It is due to the usual definition of the amplitude

$$f \propto \int \Psi^* V \Psi \, d^3 r,$$

where the integrand contains conjugate functions. However, this property is not preserved when we analytically continue the amplitude by reversing the sign of the momenta: $\bar{u}(-p_1^+)$ ceases to be the conjugate of the function $u(-p_1^+)$ (see (2.75)).

We need to have a definite prescription for the amplitudes. Hence, we would rather not treat T_{cont} as an amplitude but instead define the physical amplitude in the cross-channel to be

$$T^{e^+} = -\left(T^{e^-}\right)_{\text{cont}}. \qquad (2.78)$$

This sign is irrelevant for calculation of the cross sections.

2.5 Electron–electron scattering

We have two topologically different diagrams that contribute to electron–electron scattering in the lowest order:

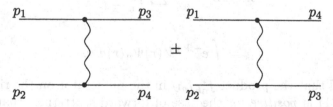

The problem arises of how to choose the relative sign of the two amplitudes? Recalling the Pauli principle, we see that the plus sign cannot be correct. In this case the amplitude would not change under the substitution $p_3 \longleftrightarrow p_4$, while it has to be antisymmetric for a spin $\frac{1}{2}$ particle. Hence, the two diagrams should be subtracted rather than added.

The question is, whether the choice of the *minus* sign can be decided upon without referring to the Pauli principle? Let us show that the *plus* sign between the amplitudes is incompatible with the relativistic interaction theory we are constructing, as it leads to an internal contradiction within our scheme.

There are two basic principles in our theory: the unitarity condition, $SS^+ = 1$, which means that the sum of all probabilities has to be unity, and that of causality. It is these two fundamental requirements that allow us to fix the sign unambiguously.

2.5.1 Connection between spin and statistics

Consider the non-relativistic scattering amplitude (compare with the discussion in Section 1.3.2)

$$f = -\frac{2m}{4\pi} \int e^{-i\mathbf{k}'\cdot\mathbf{r}'} V(r') \Psi_+(\mathbf{r}')\, d^3r'. \tag{2.79}$$

Writing the wave function $\Psi_+(\mathbf{r}')$ in terms of the Green function, we obtain

$$f = f_B + \frac{2mi}{4\pi} \int e^{-i\mathbf{k}'\cdot\mathbf{r}'} V(r') G(\mathbf{r}',\mathbf{r}) V(r) e^{i\mathbf{k}\cdot\mathbf{r}}\, d^3r\, d^3r',$$

with f_B the amplitude in the Born approximation.

In terms of the complete orthonormal set of states Ψ_n, the Green function has the form

$$G(\mathbf{r}',\mathbf{r}) = \frac{1}{i} \sum_n \frac{\Psi_n(\mathbf{r})\Psi_n^*(\mathbf{r}')}{E_n - E}, \tag{2.80}$$

and hence,

$$f = f_B + \frac{2m}{4\pi} \sum_n \frac{f_{nk}^* f_{nk'}}{E_n - E}, \tag{2.81}$$

where

$$f_{nk} = \int e^{-i\mathbf{k}\cdot\mathbf{r}} V(r) \Psi_n(\mathbf{r}) d^3r. \tag{2.82}$$

It is critical that the product $f_{nk}^* f_{nk'}$ in the numerator on the right-hand side in (2.81) is *positive* for the case of forward scattering, that is when $\mathbf{k} = \mathbf{k}'$. This positivity is in fact a result of the unitarity condition (we will consider unitarity in more detail in the next chapter). The amplitude as a function of energy has a pole at the bound state energy $E = E_n$. The positivity of the product $f_{nk}^* f_{nk}$ means that the corresponding residue is always *negative*.

We are ready now to demonstrate that the unitarity condition fixes the signs of the different diagrams unambiguously.

Consider first the case of scalar particles. We have (compare with (1.145))

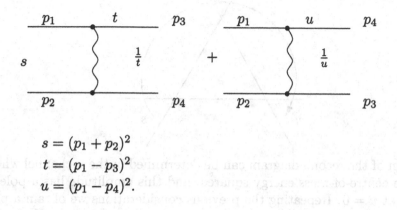

$$s = (p_1 + p_2)^2$$
$$t = (p_1 - p_3)^2$$
$$u = (p_1 - p_4)^2.$$

In the s-channel these diagrams have no singularities in energy (i.e. in s). Hence, we cannot use unitarity to fix the sign of the residue at the pole in energy.

Let us go to the t-channel, which means that we look at the diagrams from the top. The first diagram then describes the transition

$$e^-(p_1) + e^+(p_3^+) \quad \to \quad \gamma^* \quad \to \quad e^-(p_4) + e^+(p_2^+).$$

This is a second order process with a virtual γ quantum in the intermediate state. This intermediate state corresponds to the sum in (2.81). There is a pole at $t = 0$ (t in this channel is the centre-of-mass energy squared) which corresponds to the e^+e^- pair annihilation into a real photon, the intermediate state with energy $E_n = 0$ ($m_\gamma = 0$).

Now, by examining the sign of the residue at the pole at $t = 0$ and comparing with that dictated by the unitarity condition, we determine the sign of the first diagram (the second one has no singularities in the t-channel). In the centre-of-mass frame $E_1 = E_3$, $E_2 = E_4$, and in the near-to-forward scattering case, $\mathbf{p}_1 \simeq \mathbf{p}_4$, the numerator of the diagram becomes

$$(p_1 - p_3^+)_\mu (p_4 - p_2^+)_\mu \simeq -4\mathbf{p}_1^2.$$

Thus, the residue at the pole is *negative* and therefore this diagram should

2 Particles with spin $\frac{1}{2}$

enter the scattering amplitude with a plus sign.

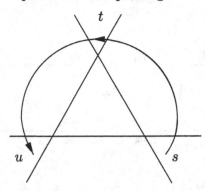

The sign of the second diagram can be determined in the u-channel where u is the centre-of-mass energy squared, and this amplitude has a pole in energy at $u = 0$. Repeating the previous considerations we obtain a plus sign for the second diagram. Hence, for bosons (spin zero particles) both diagrams should carry a plus sign (compare with (1.145) and (1.146)).

Now consider the fermions (spin $\frac{1}{2}$ particles)

$$
T^{(s)} = \pm\, e^2\, [\bar{u}(p_3)\gamma_\mu u(p_1)]\, \frac{1}{t}\, [\bar{u}(p_4)\gamma_\mu u(p_2)]
$$
$$
\pm\, e^2\, [\bar{u}(p_4)\gamma_\mu u(p_1)]\, \frac{1}{u}\, [\bar{u}(p_3)\gamma_\mu u(p_2)] . \tag{2.84}
$$

As in the previous case, we find the sign of the first amplitude by going to the t-channel. To do this, we carry out the replacement

$$
p_3 = -p_3^+, \qquad p_2 = -p_2^+ . \tag{2.85}
$$

Since $\bar{u}(-p_3^+) = -\bar{v}(p_3^+)$, and bearing in mind that the physical t-channel amplitude $T^{(t)}$ differs by sign from the analytically continued amplitude, $T^{(t)} = -T^{(s)}_{\text{cont}}$ (see (2.78)), we obtain

$$
T^{(t)} = \pm\, e^2\, \left[\bar{v}(p_3^+)\gamma_\mu u(p_1)\right]\, \frac{1}{t}\, \left[\bar{u}(p_4)\gamma_\mu v(p_2^+)\right]
$$
$$
\pm\, e^2\, [\bar{u}(p_4)\gamma_\mu u(p_1)]\, \frac{1}{u}\, \left[\bar{v}(p_3^+)\gamma_\mu v(p_2^+)\right] . \tag{2.86}
$$

We have to choose the *plus* sign for the first term in order to obtain a *negative* residue in $T^{(t)}$ at $t = 0$. Indeed, only the spatial components of currents survive in the annihilation diagram near the pole, and the currents themselves are complex conjugate to each other

$$(\bar{u}(p_4)\gamma_\mu v(p_2^+))^+ = (u^+\gamma_0\gamma_\mu v)^+ = v^+(\gamma_0\gamma_\mu)^+ u$$
$$= v^+\gamma_\mu^+\gamma_0^+ u = \bar{v}(p_2^+)\gamma_\mu u(p_4)$$

(here we used the hermiticity of γ_0 and the anti-hermiticity of γ_i). Thus, the product of the currents at $p_1 \simeq p_4$ turns out to be negative, and the residue in $T^{(t)}$ is also negative if the first diagram in (2.83) carries the plus sign.

Let us figure out the sign of the second diagram, that is the sign of the second term in (2.84). In the t-channel the second graph describes e^-e^+ scattering, and its sign may be fixed by comparison with the non-relativistic limit.

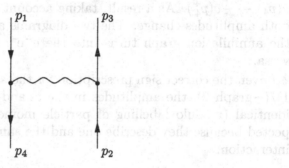

Non-relativistic amplitudes for particle–particle and antiparticle–particle scattering should have opposite signs in accordance with the signs of the respective non-relativistic potentials. In other words, the sign of the amplitude corresponding to the second diagram in (2.83) in the t-channel (e^-e^+ scattering) should be opposite to that of the first diagram in the s-channel (e^-e^-), at least for small energies and small momentum transfers. However, the amplitudes of photon emission by a particle and an antiparticle are the same, since, as we have seen, the minus sign in the relation $\bar{u}(-p_3^+)\gamma_\mu u(-p_2^+) = -\bar{v}(p_3^+)\gamma_\mu v(p_2^+)$ is compensated by the minus due to $T^{(t)} = -T_{\text{cont}}^{(s)}$. Hence, to preserve the correspondence with the non-relativistic theory, the second diagram in (2.83) should carry a minus sign, opposite to the first one.

We can also establish the sign of the second term in (2.84) without appealing to the non-relativistic limit. Let us continue the amplitude to the u-channel, where the second diagram describes electron–positron annihilation, and should therefore be positive due to unitarity (compare with the t-channel consideration of the first diagram above).

We go to the u-channel starting from the t-channel and substitute

$$p_4 = -p_4^+, \qquad p_3^+ = -p_3^{++}.$$

In the process of the $s \to t \to u$ transition we changed the sign of the vector p_3 twice: $p_3 \to -p_3^+ \to p_3^{++}$. Each time the sign of the four-momentum changes, $p \to -p$, the Dirac spinor is multiplied by i (see (2.39)) so after two substitutions it acquires the minus sign. At the same time, the sign due to $\bar{u}(p_4^+)\gamma_\mu u(p_1) = -\bar{v}(p_4^+)\gamma_\mu u(p_1)$ gets compensated by that from $T^{(u)} = -T^{(t)}_{\text{cont}}$. Therefore, to have a positive-sign expression for the annihilation amplitude in the u-channel we have to supply the second term in (2.84) with a negative sign, confirming the result we have obtained from the correspondence with non-relativistic scattering.

In the course of the $s \to t$ transition both diagrams in (2.83) retained their signs (+1 for the first term, -1 for the second one). Going from the t- to the u-channel (unlike the $s \to t$ case, $\bar{u}(p_3) \to -\bar{v}(p_3^+)$, $u(p_2) \to +v(p_2^+)$) we now continue *two conjugated* spinors, $\bar{v}(p_3^+) \to -\bar{u}(p_3^{++})$ and $\bar{u}(p_4) \to -\bar{v}(p_4^+)$. As a result, taking account of $T = -T_{\text{cont}}$, the signs of both amplitudes change. The two diagrams also interchange their rôles: the annihilation graph turns into the e^-e^+ scattering graph, and vice versa.

Given the correct sign prescription for the s-channel diagrams (+graph 1)/($-$graph 2), the amplitudes in the t- and u-channels turn out to be identical (modulo labelling of particle momenta). This was to be expected because they describe one and the same physical process of e^-e^+ interaction:

s-channel: $+$ scattering (e^-e^-) $-$ scattering (e^-e^-)
t-channel: $+$ annihilation (e^-e^+) $-$ scattering (e^-e^+)
u-channel: $-$ scattering (e^-e^+) $+$ annihilation (e^-e^+)

Thus, we have come to the conclusion that the annihilation and scattering diagrams in (2.83) should have opposite signs. Hence, the scattering amplitude in the s-channel must be antisymmetric with respect to the interchange of momenta of the initial (or of the final) particles, i.e. the electrons have to obey Fermi–Dirac statistics.

Let us remark on a subtlety concerning the *overall* sign of the interaction amplitude. If we start from the s-channel and then return to it via the t- and u-channels, the sign of the amplitude changes. However, unlike the case of the t- and u-channels where different particles interact, in the s-channel the interaction takes place between identical particles, and the overall sign is unimportant. Indeed, the unitarity condition determines only the sign for the amplitude of *forward* scattering. For identical particles, however, the processes of forward and backward scattering are the same, and only scattering into one hemisphere makes sense. So, the line $t = 0$ in the Mandelstam plane corresponds to forward scattering for the diagram with one sign, and the line $u = 0$ corresponds to forward scattering for the diagram with the other sign. In other words, the overall sign of the s-channel amplitude determines the very notion of forward scattering.

This is already true in the non-relativistic theory. Consider non-relativistic scattering of identical spin $\frac{1}{2}$ particles. The initial and final wave functions have the form

$$\Psi_a = e^{i\mathbf{p}_1 \cdot \mathbf{r}_1} e^{i\mathbf{p}_2 \cdot \mathbf{r}_2} - e^{i\mathbf{p}_1 \cdot \mathbf{r}_2} e^{i\mathbf{p}_2 \cdot \mathbf{r}_1},$$
$$\Psi_b = e^{i\mathbf{p}_3 \cdot \mathbf{r}_1} e^{i\mathbf{p}_4 \cdot \mathbf{r}_2} - e^{i\mathbf{p}_3 \cdot \mathbf{r}_2} e^{i\mathbf{p}_4 \cdot \mathbf{r}_1},$$

and the scattering amplitude

$$f_{ab} \propto \int \Psi_b^* V \Psi_a$$

has a definite sign (provided the potential has a definite sign) only if $\Psi_a = \Psi_b$, i.e. if $p_3 = p_1$, $p_2 = p_4$. Scattering at an angle $\theta > \pi/2$ is equivalent to transposing the final electrons, and the amplitude changes sign.

Let us see what will happen if, after obtaining an amplitude with sign opposite to the original one in the s-channel, we once more carry out an analytic continuation into the t- and u-channels. Obviously, the amplitudes we get in the t- and u-channels will also be of opposite sign, i.e. $T = T_{\text{cont}}$, unlike what we obtained above. Hence, the relation between the continued amplitude $e^-e^- \to e^-e^-$ and the amplitude $e^+e^- \to e^+e^-$ depends on the continuation path. This non-uniqueness is due to the uncertainty of the sign of the amplitude $e^-e^- \to e^-e^-$.

We have obtained a remarkable result here: a connection between spin and statistics. We have derived this connection from very general considerations, using the unitarity condition and the fact that an arbitrary amplitude can be obtained via analytic continuation. The latter reflects the analyticity of the amplitude which, as we will show, is connected with the causality. This means that two fundamental conditions, namely unitarity and causality, are sufficient for the determination of the signs of the amplitudes. Thus, in our theory the experimentally established Pauli principle is satisfied automatically.

2.5.2 Electron charge

In the region of small scattering angles our amplitude is simply the usual amplitude of Coulomb scattering. Having this in mind, we can show that the coupling constant e in (2.84) is just the electric charge. Small scattering angles correspond to $p_3 \simeq p_1$, $p_2 \simeq p_4$, i.e. $t \simeq 0$. The first amplitude is proportional to $1/t$, so that we can neglect the second (exchange) amplitude which remains finite in the limit $t \to 0$.

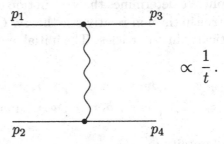

Let us calculate $\bar{u}(p_3)\gamma_\mu u(p_1)$ for $p_3 \simeq p_1$. Using $\hat{p}_1 + m = \hat{p}_1 - m + 2m$, $(\hat{p}_1 - m)u = 0$, and

$$\gamma_\mu \hat{p}_1 = -\hat{p}_1 \gamma_\mu + 2p_{1\mu},$$

we can write

$$\bar{u}(p_3)\gamma_\mu u(p_1) \simeq \bar{u}(p_1)\gamma_\mu u(p_1) \simeq \bar{u}(p_1)\gamma_\mu \frac{m + \hat{p}_1}{2m} u(p_1)$$

$$= \bar{u}(p_1)\frac{m - \hat{p}_1}{2m}\gamma_\mu u(p_1) + 2p_{1\mu}\frac{\bar{u}(p_1)u(p_1)}{2m} = 2p_{1\mu}. \tag{2.87}$$

Similarly,

$$\bar{u}(p_4)\gamma_\mu u(p_2) \simeq 2p_{2\mu}. \tag{2.88}$$

Then in the non-relativistic limit we have

$$T = \frac{e^2}{t} 2p_{1\mu} \cdot 2p_{2\mu} \simeq \frac{4e^2 m^2}{t}. \tag{2.89}$$

This is the usual Coulomb scattering amplitude, which coincides with (1.147) for spinless particles. In other words, the spin of the electron plays no rôle at small momentum transfers: spinor vertices (2.87) and (2.88) coincide with the electrodynamic vertices for scalar particles, $p_{1\mu} + p_{3\mu} \simeq 2p_{1\mu}$, $p_{2\mu} + p_{4\mu} \simeq 2p_{2\mu}$. We see from (2.89) that the coupling constant e is the electric charge.

2.6 The Compton effect

Consider, as usual, the simplest diagrams describing the Compton effect.

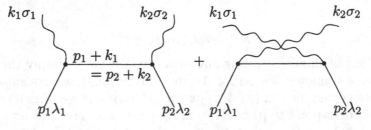

Fig. 2.1

The second of these graphs may also be drawn as

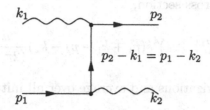

The Compton scattering amplitude can be written as

$$T = e_\nu^{\sigma_2} \bar{u}^{\lambda_2}(p_2) M_{\nu\mu} u^{\lambda_1}(p_1) e_\mu^{\sigma_1}, \tag{2.90}$$

where

$$
\begin{aligned}
M_{\nu\mu} &= e^2 \left[\gamma_\nu \frac{1}{m - \hat{p}_1 - \hat{k}_1} \gamma_\mu + \gamma_\mu \frac{1}{m - \hat{p}_1 + \hat{k}_2} \gamma_\nu \right] \\
&= e^2 \left[\frac{\gamma_\nu(\hat{p}_1 + \hat{k}_1 + m)\gamma_\mu}{m^2 - (p_1 + k_1)^2} + \frac{\gamma_\mu(\hat{p}_1 - \hat{k}_2 + m)\gamma_\nu}{m^2 - (p_1 - k_2)^2} \right],
\end{aligned} \tag{2.91}
$$

and

$$(p_1 + k_1)^2 = s, \qquad (p_1 - k_2)^2 = u.$$

To show that the current conservation conditions

$$k_{2\nu} M_{\nu\mu} = 0, \qquad k_{1\mu} M_{\nu\mu} = 0 \tag{2.92}$$

are satisfied automatically, let us calculate

$$k_{2\nu} M_{\nu\mu} = e^2 \left[\hat{k}_2 \frac{1}{m - \hat{p}_2 - \hat{k}_2} \gamma_\mu + \gamma_\mu \frac{1}{m - \hat{p}_1 + \hat{k}_2} \hat{k}_2 \right].$$

The amplitude $M_{\nu\mu}$ enters only between the on-mass-shell spinors, hence, we can add $\hat{p}_2 - m$ and $\hat{p}_1 - m$ in the numerators. Due to the Dirac equation the amplitude will not change. So,

$$
\begin{aligned}
k_{2\nu} M_{\nu\mu} &= e^2 \left[(\hat{k}_2 + \hat{p}_2 - m) \frac{1}{m - \hat{p}_2 - \hat{k}_2} \gamma_\mu + \gamma_\mu \frac{1}{m - \hat{p}_1 + \hat{k}_2} (\hat{k}_2 + m - \hat{p}_1) \right] \\
&= e^2(-\gamma_\mu + \gamma_\mu) = 0.
\end{aligned}
$$

Similarly, we can prove the second identity in (2.92).

Calculation of the amplitude is difficult because of the large number of spin variables. To avoid complications, let us consider the simplest experimental situation and calculate the total cross section for scattering into all possible electron and photon polarizations in the final state, for the case when the incident beams are unpolarized. This means that we have to sum the cross section,

$$d\sigma = \frac{1}{J} \left| T^{\sigma_2 \sigma_1}_{\lambda_2 \lambda_1} \right|^2 (2\pi)^4 \delta(p_1 + k_1 - p_2 - k_2) \frac{d^3 p_2 d^3 k_2}{(2\pi)^6 2k_{20} 2p_{20}}, \qquad (2.93)$$

over all final polarizations and average over all initial polarizations:

$$\sum \equiv \frac{1}{4} \sum_{\substack{\sigma_2 \sigma_1 \\ \lambda_2 \lambda_1}} \left| T^{\sigma_2 \sigma_1}_{\lambda_2 \lambda_1} \right|^2 = \frac{1}{4} \sum e^{\sigma_1}_\mu e^{\sigma_2}_\nu e^{\sigma_1 *}_{\mu'} e^{\sigma_2 *}_{\nu'}$$

$$\times \left(\bar{u}^{\lambda_2}(p_2) M_{\nu\mu} u^{\lambda_1}(p_1) \right) \left(\bar{u}^{\lambda_2}(p_2) M_{\nu'\mu'} u^{\lambda_1}(p_1) \right)^\dagger . \qquad (2.94)$$

Summation goes only over two transverse photon polarizations $\sigma = 1, 2$. Conservation of current, however, allows us to perform summation over all four polarizations, since the extra polarizations will not contribute to the sum due to (2.92). We then effectively have (see discussion after (1.88) in Section 1.5.4)

$$\sum_{\sigma_1 = 1,2} e^{\sigma_1}_\mu e^{\sigma_1 *}_{\mu'} \rightarrow \sum_{\sigma_1 = 0}^{3} e^{\sigma_1}_\mu e^{\sigma_1 *}_{\mu'} = -g_{\mu\mu'}.$$

Introducing

$$\bar{M}_{\mu\nu} = \gamma_0 M^\dagger_{\mu\nu} \gamma_0 \qquad (2.95)$$

and using the identity $\gamma_0 \gamma_0 = 1$, we can write (2.94) as

$$\begin{aligned}
\sum &= \frac{1}{4} \sum_{\lambda_1 \lambda_2} (\bar{u}^{\lambda_2} M_{\nu\mu} u^{\lambda_1})(\bar{u}^{\lambda_2} M_{\nu\mu} u^{\lambda_1})^\dagger \\
&= \frac{1}{4} \sum_{\lambda_1 \lambda_2} (\bar{u}^{\lambda_2} M_{\nu\mu} u^{\lambda_1})(u^{\lambda_1 *} M^\dagger_{\nu\mu} \gamma_0 u^{\lambda_2}) \\
&= \frac{1}{4} \sum_{\lambda_1 \lambda_2} (\bar{u}^{\lambda_2} M_{\nu\mu} u^{\lambda_1})(\bar{u}^{\lambda_1} \gamma_0 M^\dagger_{\nu\mu} \gamma_0 u^{\lambda_2}) \\
&= \frac{1}{4} \sum_{\lambda_1 \lambda_2} (\bar{u}^{\lambda_2} M_{\nu\mu} u^{\lambda_1})(\bar{u}^{\lambda_1} \bar{M}_{\nu\mu} u^{\lambda_2}).
\end{aligned} \qquad (2.96)$$

From the explicit form (2.91) of the amplitude $M_{\nu\mu}$ it follows that

$$\bar{M}_{\nu\mu} = M_{\mu\nu}. \qquad (2.97)$$

(To verify, recall $\{\gamma_0 \gamma_i\} = 0$, $\gamma_0^\dagger = \gamma_0$, $\gamma_i^\dagger = -\gamma_i$.)

Now we are ready to sum over fermion polarizations λ_1, λ_2. Let us write (2.96) in the matrix form

$$\Sigma = \frac{1}{4} \sum_{\lambda_1 \lambda_2} \bar{u}_\alpha^{\lambda_2}(p_2)(M_{\nu\mu})_{\alpha\beta} u_\beta^{\lambda_1}(p_1) \bar{u}_\gamma^{\lambda_1}(p_1)(M_{\mu\nu})_{\gamma\delta} u_\delta^{\lambda_2}(p_2)$$

$$= \frac{1}{4} \sum_{\alpha\beta\gamma\delta} (M_{\nu\mu})_{\alpha\beta}(\hat{p}_1 + m)_{\beta\gamma}(M_{\mu\nu})_{\gamma\delta}(\hat{p}_2 + m)_{\delta\alpha} \qquad (2.98)$$

$$= \frac{1}{4} \operatorname{Tr}[M_{\nu\mu}(\hat{p}_1 + m)M_{\mu\nu}(\hat{p}_2 + m)],$$

where we have used the identity

$$\sum_\lambda u_\alpha^\lambda(p)\bar{u}_\beta^\lambda(p) = (\hat{p} + m)_{\alpha\beta}.$$

Thus, we have reduced the summation over the fermion polarizations to calculation of the trace of a matrix.

It is convenient to write the phase volumes in the form

$$\frac{d^3 p_2}{2p_{20}} = d^4 p_2 \delta_+(p_2^2 - m^2)$$

$$\frac{d^3 k_2}{2k_{20}} = d^4 k_2 \delta_+(k_2^2).$$

The expression for the cross section then becomes

$$d\sigma = \frac{1}{4J} \operatorname{Tr}\left[(\hat{p}_1 + m)M_{\mu\nu}(\hat{p}_2 + m)M_{\nu\mu}\right]$$

$$\times \delta_+(p_2^2 - m^2)\delta_+(k_2^2)(2\pi)^4 \delta(p_1 + k_1 - p_2 - k_2)\frac{d^4 k_2 d^4 p_2}{(2\pi)^6}. \qquad (2.99)$$

As in the case of scalar particles, the calculation of the cross section (2.99) can be described graphically. Consider the first diagram. It has to be multiplied by its Hermitian conjugate which corresponds to the interchange of k_1, p_1 and k_2, p_2:

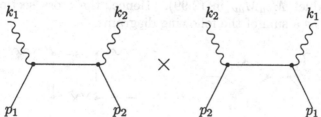

Instead of this symbolic product we draw

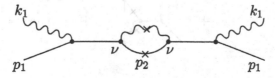

This new diagram is convenient from a purely technical point of view: it shows that the cross section may be calculated by the same rules as the amplitude. The difference is that for the lines marked by $\times$ the denominators in the propagators $1/k_2^2$ and $(m + \hat{p}_2)/(m^2 - p_2^2) = 1/(m - \hat{p}_2)$ should be substituted by the δ-functions, i.e. real particles correspond to such lines. The factor $(\hat{p}_2 + m) = \sum u(p_2)\bar{u}(p_2)$ in the numerator describes the sum over polarizations of the final electron while $(\hat{p}_1 + m) = \sum u(p_1)\bar{u}(p_1)$ arises from averaging over the initial polarization states. Apart from the substitution $1/(m^2 - p^2) \to \delta_+(m^2 - p^2)$ for the final state particles, the only difference between the diagrams for the amplitude and the cross section is an extra factor $(2\pi)^2$ in the latter case.

Similarly, squaring the second graph in Fig. 2.1 we obtain

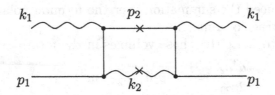

In addition, interference terms arise from the multiplication of different diagrams in Fig. 2.1:

All these contributions can easily be obtained from the explicit expression for the product $M_{\nu\mu}M_{\mu\nu}$ in (2.99). Hence, the cross section may be represented as a sum of the following diagrams:

Let us now calculate the trace. For the first diagram we have

$$f(s, u) = \frac{1}{4(m^2 - s)^2} \, \text{Tr} \left[(\hat{p}_1 + \hat{k}_1 + m)\gamma_\mu(\hat{p}_1 + m)\gamma_\mu \right.$$
$$\left. \times (\hat{p}_1 + \hat{k}_1 + m)\gamma_\nu(\hat{p}_2 + m)\gamma_\nu \right]. \tag{2.100}$$

We have inserted into (2.98) the first term of $M_{\nu\mu}$ given in (2.91) that corresponds to the first diagram, and shifted γ_ν from the beginning to the end of the expression for the trace, using its cyclic invariance.

Some useful auxiliary formulae which simplify further calculations are due:

$$\frac{1}{4} \, \text{Tr}(\gamma_\mu \gamma_\nu) = g_{\mu\nu}, \tag{2.101}$$

$$\frac{1}{4} \, \text{Tr}(\gamma_{\mu_1} \gamma_{\mu_2} \gamma_{\mu_3} \gamma_{\mu_4}) = g_{\mu_1\mu_2} g_{\mu_3\mu_4} + g_{\mu_2\mu_3} g_{\mu_1\mu_4} - g_{\mu_1\mu_3} g_{\mu_2\mu_4}. \tag{2.102}$$

Trace of the product of an *odd* number of γ_μ-matrices equals zero.

Let us show, for example, how to obtain (2.101). We have

$$\frac{1}{4} \, \text{Tr}(\gamma_\mu \gamma_\nu) = \frac{1}{4} \, \text{Tr}(\gamma_\nu \gamma_\mu).$$

On the other hand,

$$\frac{1}{4} \, \text{Tr}(\gamma_\mu \gamma_\nu) = -\frac{1}{4} \, \text{Tr}(\gamma_\nu \gamma_\mu) + \frac{2}{4} \, \text{Tr}(I) \, g_{\mu\nu}.$$

Subtracting the two equalities we get (2.101). It is also easy to check (2.102).

Applying the commutation relations for the γ-matrices, it is straightforward to derive the following useful relations involving arbitrary matrices A, B, and C:

$$\gamma_\mu \hat{C} \gamma_\mu = -2\hat{C}, \tag{2.103}$$
$$\gamma_\mu \hat{A}\hat{B}\hat{C} \gamma_\mu = -2\hat{C}\hat{B}\hat{A}. \tag{2.104}$$

For example, the first identity may be proved as follows:

$$\gamma_\mu \hat{C} \gamma_\mu = C_\nu \gamma_\mu \gamma_\nu \gamma_\mu = -C_\nu \gamma_\nu \gamma_\mu \gamma_\mu + 2C_\nu g_{\mu\nu} \gamma_\mu$$
$$= -4C_\nu \gamma_\nu + 2C_\nu \gamma_\nu = -2\hat{C}, \qquad (\gamma_\mu \gamma_\mu = 4).$$

Applying (2.103), we get

$$f(s, u) = \frac{1}{4} \frac{4}{(m^2 - s)^2}$$
$$\times \text{Tr} \left[(\hat{p}_1 + \hat{k}_1 + m)(2m - \hat{p}_1)(\hat{p}_1 + \hat{k}_1 + m)(2m - \hat{p}_2) \right]. \tag{2.105}$$

Only the terms in (2.105) that contain products of even numbers of γ-matrices (0,2,4) give non-vanishing contributions. Using (2.101) and (2.102), we obtain

$$f(s,u) = \frac{4}{(m^2-s)^2}\Big\{ 4m^4 - 2m^2 p_2(p_1+k_1) + m^2 p_1 p_2 - 2m^2 p_1(p_1+k_1)$$

$$- 2m^2 p_2(p_1+k_1) + 4m^2(p_1+k_1)^2 - 2m^2 p_1(p_1+k_1)$$

$$+ \frac{1}{4}\operatorname{Tr}\Big[(\hat{p}_1 + \hat{k}_1)\hat{p}_2(\hat{p}_1 + \hat{k}_1)\hat{p}_1\Big]\Big\}$$

$$= \frac{4}{(m^2-s)^2}\Big\{ 4m^4 - 4m^2(p_1+p_2)(p_1+k_1) + m^2 p_1 p_2$$

$$+ 4m^2 s + \frac{1}{4}\operatorname{Tr}\Big[(\hat{p}_1 + \hat{k}_1)\hat{p}_2(\hat{p}_1 + \hat{k}_1)\hat{p}_1\Big]\Big\}.$$

Note that

$$(p_1 + p_2)(p_1 + k_1) = s + m^2,$$

since $p_1 + k_1 = p_2 + k_2$ and

$$2p_1(p_1+k_1) = 2m^2 + 2p_1 k_1 = m^2 + (m^2 + 2p_1 k_1) = m^2 + (p_1+k_1)^2 \equiv m^2 + s.$$

This leads to

$$f(s,u) = \frac{1}{(m^2-s)^2}\Big\{ 4m^2 p_1 p_2 + \operatorname{Tr}\Big[(\hat{p}_1 + \hat{k}_1)\hat{p}_2(\hat{p}_1 + \hat{k}_1)\hat{p}_1\Big]\Big\}.$$

Calculating the remaining trace with the help of (2.102),

$$\operatorname{Tr}\Big[(\hat{p}_1 + \hat{k}_1)\hat{p}_2(\hat{p}_2 + \hat{k}_2)\hat{p}_1\Big] = 2(s + m^2)^2 - 4s\, p_1 p_2,$$

we arrive at

$$f(s,u) = \frac{2}{(m^2-s)^2}\Big[(s+m^2)^2 - 2p_1 p_2(s - m^2)\Big].$$

We can write $2p_1 p_2$ in terms of the invariant variables:

$$t = (p_1 - p_2)^2 = 2m^2 - 2p_1 p_2,$$

so that

$$2p_1 p_2 = 2m^2 - t = s + u. \qquad (s + t + u = 2m^2)$$

Finally, this gives

$$f(s,u) = \frac{2}{(m^2-s)^2}\Big[(s+m^2)^2 - (s+u)(s-m^2)\Big]$$

$$= \frac{2}{(m^2-s)^2}\Big[4m^4 - (u-m^2)(s-m^2) + 2m^2(s-m^2)\Big]. \qquad (2.106)$$

The second form of the answer is better suited for exploiting the $s \leftrightarrow u$ symmetry: replacing s by u in (2.106), we get the expression for the second diagram for the cross section (the square of the second term in the amplitude (2.91)). Indeed, this term can be obtained from the first one by the substitution $k_1 \to -k_2$ and $\mu \leftrightarrow \nu$. Renaming the vector indices does not affect the result. Substituting $-k_2$ for k_1 results in $s = (p_1 + k_1)^2 \to (p_1 - k_2)^2 = u$. Hence, the contribution of the second diagram is just $f(u, s)$.

The traces of the interference terms (we call them $g(s, u)$ and $g(u, s)$) may be calculated in a similar way:

$$g(s, u) = \frac{2m^2}{(m^2 - s)(m^2 - u)} \left[4m^2 + s - m^2 + u - m^2\right]. \qquad (2.107)$$

We have

$$\frac{1}{4} \operatorname{Tr} \left[(\hat{p}_1 + m) M_{\nu\mu} (\hat{p}_2 + m) M_{\mu\nu}\right]$$

$$= e^4 \left[f(s, u) + f(u, s) + g(u, s) + g(s, u)\right]. \qquad (2.108)$$

Taking into account (2.108), the cross section (2.99) can be written as

$$d\sigma = \frac{e^4}{J} [f(s, u) + f(u, s) + g(u, s) + g(s, u)]$$

$$\times \delta_+(p_2^2 - m^2)\delta_+(k_2^2)\delta(p_1 + k_1 - p_2 - k_2)\frac{d^4 k_2 d^4 p_2}{(2\pi)^2}, \qquad (2.109)$$

where the invariant flux is the same as in (1.138),

$$J = 4p_{01}k_{01}\, j, \qquad (2.110)$$

with the relative flux j given by the individual fluxes of the colliding particles as in (1.137). Explicitly, we have

$$j = |\bar{\Psi}(p_1, x)\boldsymbol{\gamma}\Psi(p_1, x)| + \frac{|\mathbf{k}|}{k_0} = \frac{|\bar{u}(p_1)\boldsymbol{\gamma}u(p_1)|}{2p_{10}} + \frac{|\mathbf{k}_1|}{k_{10}} = \frac{|\mathbf{p}_1|}{p_{10}} + \frac{|\mathbf{k}_1|}{k_{10}}.$$

We still have to calculate the phase volume. Let us first integrate (2.109) over p_2 with the help of the δ-function. We get

$$d\sigma = \frac{e^4}{J} [f(s, u) + f(u, s) + g(u, s) + g(s, u)]$$

$$\times \delta_+((p - k_2)^2 - m^2)\delta_+(k_2^2)\frac{d^4 k_2}{(2\pi)^2}, \qquad (2.111)$$

where $p = p_1 + k_1$, and

$$(p - k_2)^2 = p^2 - 2pk_2 + k_2^2 = s - 2pk_2.$$

In the centre-of-mass frame we have

$$\delta_+((p - k_2)^2 - m^2) = \delta(s - 2p_0 k_{20} - m^2) = \delta(s - 2\sqrt{s}k_{20} - m^2). \quad (2.112)$$

The corresponding invariant flux is

$$\mathcal{J} = 4(|\mathbf{p}_1|k_{10} + |\mathbf{k}_1|p_{10}) = 4|\mathbf{k}_1|(k_{10} + p_{10}) = 4|\mathbf{k}_1|\sqrt{s}. \quad (2.113)$$

Let us introduce an invariant phase volume element

$$d\Gamma \equiv \frac{1}{\mathcal{J}}\delta_+((p - k_2)^2 - m^2)\delta_+(k_2^2)\frac{d^4 k_2}{(2\pi)^2}. \quad (2.114)$$

Taking into account (2.112) and (2.113),

$$d\Gamma = \frac{1}{4|\mathbf{k}_1|\sqrt{s}}\delta_+\left(2\sqrt{s}\left[\frac{\sqrt{s}}{2} - \frac{m^2}{2\sqrt{s}} - k_{20}\right]\right)\delta_+(k_{20}^2 - \mathbf{k}_2^2)\frac{d^4 k_2}{(2\pi)^2}.$$

Integrating over k_{20} with the help of the first δ-function, we have

$$d\Gamma = \frac{1}{8|\mathbf{k}_1|s}\delta_+\left(\left[\frac{s - m^2}{2\sqrt{s}}\right]^2 - \mathbf{k}_2^2\right)\frac{d^3 k_2}{(2\pi)^2}.$$

Introducing spherical coordinates

$$d^3 k_2 = \mathbf{k}_2^2 d|\mathbf{k}_2|d\Omega = \frac{|\mathbf{k}_2|}{2}d\mathbf{k}_2^2 d\Omega,$$

we integrate over $\mathbf{k}_2^2$:

$$d\Gamma = \frac{1}{16s}\frac{|\mathbf{k}_2|}{|\mathbf{k}_1|}\frac{d\Omega}{(2\pi)^2}. \quad (2.115)$$

The δ-function gives us the photon momentum:

$$|\mathbf{k}_2| = \frac{s - m^2}{2\sqrt{s}}. \quad (2.116)$$

In the centre-of-mass frame, however, $|\mathbf{k}_1| = |\mathbf{k}_2| \equiv k$. The initial- and final-state momenta cancel in the ratio, and we have

$$d\Gamma = \frac{1}{16s}\frac{d\Omega}{(2\pi)^2}. \quad (2.117)$$

Let us represent $d\Omega$ in terms of the invariant variables

$$t = -2k^2(1 - \cos\theta); \qquad dt = 2k^2 d(\cos\theta).$$

Since $d\Omega = d(\cos\theta)d\varphi = 2\pi d(\cos\theta)$, we have

$$d\Omega = 2\pi \frac{dt}{2k^2}, \qquad d\Gamma = \frac{1}{16s}\frac{dt}{2k^2}\frac{1}{2\pi}.$$

We take now k^2 from (2.116) to finally obtain

$$d\Gamma = \frac{1}{16\pi}\frac{dt}{(m^2 - s)^2}. \qquad (2.118)$$

This is a standard procedure for the calculation of phase volumes.

We are now ready to write the expression for the cross section. Taking the phase volume (2.118) and substituting (2.106) and (2.107) into (2.111), we obtain the final expression for the cross section of elastic electron–photon scattering:

$$
\begin{aligned}
d\sigma = \frac{e^4}{16\pi}\frac{dt}{(m^2 - s)^2}\, 8 &\left[\left(\frac{m^2}{s - m^2} + \frac{m^2}{u - m^2}\right)^2\right.\\
&\left.+ \left(\frac{m^2}{s - m^2} + \frac{m^2}{u - m^2}\right) - \frac{1}{4}\left(\frac{u - m^2}{s - m^2} + \frac{s - m^2}{u - m^2}\right)\right].
\end{aligned}
\qquad (2.119)
$$

This is the well-known Klein–Nishina formula.

2.6.1 Compton scattering at small energies

Consider Compton scattering in the laboratory frame where the initial electron is at rest. In this case

$$s = (p_1 + k_1)^2 = (k_{10} + m)^2 - \mathbf{k}_1^2 = m^2 + 2mk_{10}.$$

Denote $k_{10} = \omega$, $k_{20} = \omega'$. Then

$$s = m^2 + 2m\omega$$

and, similarly,

$$u = (p_1 - k_2)^2 = m^2 - 2m\omega'.$$

For the momentum transfer between the photons we have

$$t = (k_1 - k_2)^2 = -2k_1 k_2 = -2\omega\omega' + 2\omega\omega'\cos\theta = -2\omega\omega'(1 - \cos\theta).$$

On the other hand,

$$t = (p_1 - p_2)^2 = 2m^2 - 2mp_{20} = 2m(m - p_{20}) = 2m(\omega' - \omega),$$

i.e.

$$2\omega\omega'(1 - \cos\theta) = 2m(\omega - \omega').$$

Hence, the photon energy change after the scattering (the Compton frequency shift) is

$$m\left(\frac{1}{\omega'} - \frac{1}{\omega}\right) = 1 - \cos\theta. \qquad (2.120)$$

Let us consider separate terms on the right-hand side of (2.119) in the laboratory frame:

$$\frac{m^2}{s - m^2} = \frac{m}{2\omega}, \qquad \frac{m^2}{u - m^2} = -\frac{m}{2\omega'}.$$

Summing them, we get

$$\frac{m^2}{s - m^2} + \frac{m^2}{u - m^2} = \frac{m}{2}\left(\frac{1}{\omega} - \frac{1}{\omega'}\right) = -\frac{1}{2}(1 - \cos\theta).$$

Similarly,

$$\frac{u - m^2}{s - m^2} = -\frac{\omega'}{\omega}$$

and

$$dt = 2\omega\omega' d(\cos\theta) - 2\omega(1 - \cos\theta)\,d\omega' = -\frac{\omega}{m}(1 - \cos\theta)dt + 2\omega\omega' d(\cos\theta),$$

which, with the help of (2.120), leads to

$$dt = 2\omega'^2\, d(\cos\theta).$$

Inserting these expressions in (2.119), we obtain

$$
\begin{aligned}
d\sigma &= \frac{e^4}{4\pi m^2}\left(\frac{\omega'}{\omega}\right)^2\left[\frac{1}{4}\left(\frac{\omega'}{\omega} + \frac{\omega}{\omega'}\right) - \frac{1}{4}\sin^2\theta\right]d(\cos\theta) \\
&= \left(\frac{e^2}{4\pi}\right)^2\frac{1}{2m^2}\left(\frac{\omega'}{\omega}\right)^2\left[\frac{\omega'}{\omega} + \frac{\omega}{\omega'} - \sin^2\theta\right]d\Omega_{\omega'},
\end{aligned}
\qquad (2.121)
$$

where $d\Omega_{\omega'} = 2\pi d(\cos\theta)$.

For small energies $\omega \ll m$ of the initial photon (Thomson limit), from (2.120) follows $\omega'/\omega \to 1$, the cross section becomes energy-independent and we obtain the Rayleigh–Thomson formula:

$$\frac{d\sigma}{d\Omega_{\omega'}} \to \left(\frac{e^2}{4\pi}\right)^2\frac{2 - \sin^2\theta}{2m^2}. \qquad (2.122)$$

Note that $e^2/4\pi m \equiv r_e \simeq 2.8 \cdot 10^{-13}$ cm is the classical electron radius. This means that at small energies

$$\sigma_{e\gamma} \sim \pi r_e^2.$$

2.6.2 Compton scattering at high energies

Let us discuss the behaviour of the cross sections at high energies $s \gg m^2$. For this purpose it will be more convenient to use (2.119).

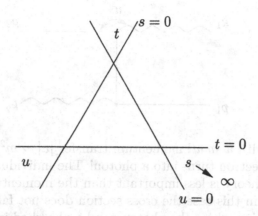

(1) Consider first the region $|t| \sim m^2 \ll s \simeq |u|$. In this case

$$d\sigma \simeq \frac{e^4}{4\pi} \frac{dt}{(m^2 - s)^2} \simeq \frac{e^4 dt}{4\pi s^2} \propto \frac{d\Omega}{s}, \tag{2.123}$$

i.e. the cross section in the region of small momentum transfer decreases rather fast with the growth of s.

(2) In the region of small u (i.e. at very large momentum transfer $|t| \simeq s$, $s + t = \mathcal{O}(m^2)$) the cross section in a unit solid angle,

$$d\sigma \simeq \frac{e^4}{8\pi} \frac{dt}{s} \frac{1}{m^2 - u} \propto \frac{d\Omega}{m^2}, \tag{2.124}$$

is larger and does not depend on s. This means that at high energies the photons in the centre-of-mass frame scatter mainly *backward*, since

$$u = (p_1 - k_2)^2 \simeq -2p_c^2(1 + \cos \theta) \simeq -\frac{s}{2}(1 + \cos \theta),$$

and finite $|u| = \mathcal{O}(m^2)$ correspond to $\pi - \theta \sim m^2/s \to 0$.

How can we explain why the photons scatter mainly at 180°? It suffices to have a look at the two diagrams in Fig. 2.1. The first one corresponds to interaction of point-like particles. Here only one partial wave contributes so that the cross section does not depend on the angle (t, u), and its size is determined by the wavelength of the intermediate virtual state:

$$\sigma < 4\pi\lambda^2 \sim \frac{4\pi}{k^2} \propto \frac{1}{s} \to 0.$$

Now look at the second diagram, the one that is responsible for the scattering peak in the backward direction. The process described by the graph

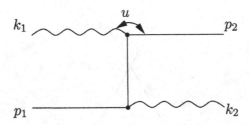

actually goes with a *small* momentum transfer $|u| \sim m^2$. In this process, however, the electron turns into a photon! The individuality of a particle in relativistic theory is less important than the momentum transfer. It is also clear why in this case the cross section does not fall with increasing energy: the region where the photon can be absorbed is now determined not by the small photon wavelength λ but by the distance between the interaction points.

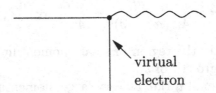

This distance can be estimated from the uncertainty relations. A virtual electron exists during the time interval

$$\Delta t \sim \frac{1}{\Delta E} \sim \frac{1}{m}$$

and propagates at a finite distance $\Delta r \sim 1/m$. This is why only the u-channel exchange process is relevant at high energies, in accordance with (2.124). The *total* cross section, however, remains small, since the backward peak where the distribution is finite,

$$\frac{d\sigma}{d\Omega} \sim \frac{e^4}{m^2},$$

is very narrow: $|d\Omega| \propto m^2/s$.

2.7 Electron–positron annihilation into two photons

$$e^+ + e^- \to 2\gamma$$

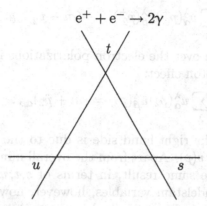

The annihilation of an electron–positron pair into two photons is the t-channel partner process of photon–electron (Compton) scattering. So, we take the diagrams for the latter,

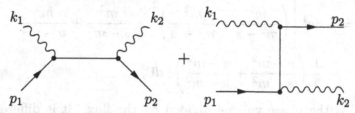

and carry out the substitutions

$$p_2 = -p_2^+, \qquad k_1 = -k_1^+. \tag{2.125}$$

The picture can be redrawn then as

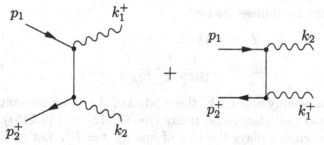

The internal parts of the diagrams do not change and are still described by the tensor $M_{\nu\mu}$, as before.

The annihilation cross section has the form

$$
d\sigma = \frac{1}{4} \sum_{\substack{\sigma_1\sigma_2 \\ \lambda_1\lambda_2}} |\bar{v}^{\lambda_2}(p_2^+) e_\nu^{\sigma_2} M_{\nu\mu} e_\mu^{\sigma_1} u^{\lambda_1}(p_1)|^2 d\Gamma
$$

$$
= -\frac{1}{4} \operatorname{Tr} \left[M_{\nu\mu}(\hat{p}_1 + m) M_{\mu\nu}(m - \hat{p}_2^+) \right] d\Gamma. \tag{2.126}
$$

We have summed here over the positron polarizations in the initial state,

$$\sum_\lambda v_\alpha^\lambda(p_2^+)\bar{v}_\beta^\lambda(p_2^+) = -(m - \hat{p}_2^+)_{\alpha\beta},$$

instead of summation over the electron polarizations in the final state in the case of the Compton effect:

$$\sum_\lambda u_\alpha^\lambda(p_2)\bar{u}_\beta^\lambda(p_2) = (m + \hat{p}_2)_{\alpha\beta}.$$

The minus sign on the right-hand side is due to the fact that $\bar{u}(-p) = -\bar{v}(p)$ while $u(-p) = v(p)$. Apart from the overall minus sign, calculation of the trace gives the same result, in terms of s, t, u, as in the case of scattering. The Mandelstam variables, however, now have a new interpretation: t is the total energy squared in the centre-of-mass frame, and $s, u < 0$ are the momentum transfers.

Thus, the cross section in invariant variables reads

$$d\sigma = -8e^4 \left[\left(\frac{m^2}{m^2 - s} + \frac{m^2}{m^2 - u} \right)^2 + \frac{m^2}{s - m^2} + \frac{m^2}{u - m^2} \right.$$
$$\left. - \frac{1}{4} \left(\frac{u - m^2}{s - m^2} + \frac{s - m^2}{u - m^2} \right) \right] d\Gamma, \qquad (2.127)$$

where $d\Gamma$ is the phase volume divided by the flux. It is different from that in the case of Compton scattering, since both the phase volume of two photons and the flux of the initial e^-e^+ differ from the phase volume and the flux for an electron and a photon.

We have already calculated the invariant flux and the two-particle phase volume in the centre-of-mass frame:

$$\mathcal{J} = 4 k_i E_c, \qquad (2.128)$$

$$d\Gamma = \frac{1}{16E_c^2} \frac{k_f}{k_i} \frac{d\Omega}{(2\pi)^2}, \qquad (2.129)$$

with E_c the total energy and k_i, k_f the moduli of the c.m. three-momenta of initial and final particles, respectively (see (2.113) and (2.115)).

In the present case, t plays the rôle of energy, $t = E_c^2$. Let us find the momenta $k_i = |\mathbf{p_1}| = |\mathbf{p}_2^+|$ and $k_f = |\mathbf{k}_1^+| = |\mathbf{k_2}|$ in terms of invariant variables. In the centre of mass of two particles, $\mathbf{p_1} = -\mathbf{p_2}$, $|\mathbf{p_1}| = |\mathbf{p_2}| \equiv \mathbf{k}$, we have

$$E_c = p_{10} + p_{20} = \sqrt{m_1^2 + k^2} + \sqrt{m_2^2 + k^2}.$$

Solving the quadratic equation for k we obtain the general expression

$$k = \frac{1}{2E_c} \sqrt{E_c^4 - 2E_c^2(m_1^2 + m_2^2) + (m_1^2 - m_2^2)^2}. \qquad (2.130)$$

In our case the initial state consists of two electrons ($m_1 = m_2 = m$), the final state of two photons ($m_1 = m_2 = 0$), and (2.130) gives

$$k_i = \frac{\sqrt{t - 4m^2}}{2}, \qquad k_f = \frac{\sqrt{t}}{2}. \tag{2.131}$$

In the cross channel of the photon–electron scattering $m_1 = m$, $m_2 = 0$, $E_c^2 = s$, and we would obtain from (2.130)

$$k_i = k_f = \frac{s - m^2}{2\sqrt{s}}$$

(compare with (2.116)). The centre-of-mass momenta coincide because for the photon–electron scattering the initial and the final state contain the same particles.

Substituting (2.131) into (2.129) we derive

$$d\Gamma = \frac{1}{16\sqrt{t(t - 4m^2)}} \frac{d\Omega}{(2\pi)^2}. \tag{2.132}$$

Writing $d\Omega$ in terms of invariant variables, we could obtain a relativistically invariant expression, valid in an arbitrary reference frame. We have done such an invariant calculation above for the Compton effect.

Let us now analyse the annihilation cross section

$$d\sigma = -\frac{e^4}{2} \frac{1}{\sqrt{t(t - 4m^2)}} \left[\left(\frac{m^2}{m^2 - s} + \frac{m^2}{m^2 - u} \right)^2 \right.$$
$$\left. + \frac{m^2}{s - m^2} + \frac{m^2}{u - m^2} - \frac{1}{4} \left(\frac{u - m^2}{s - m^2} + \frac{s - m^2}{u - m^2} \right) \right] \frac{d\Omega}{(2\pi)^2}. \tag{2.133}$$

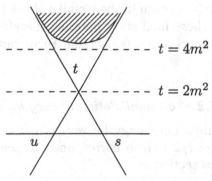

The physical region for this process is determined by the condition $t > 4m^2$. The cross section (2.133) is obviously $s \leftrightarrow u$ symmetric, as this transformation corresponds to interchanging the identical final state photons. In terms of the angle θ between the directions of the initial e^+e^-

and the final photons in the c.m. frame, the momentum transfer variables read

$$m^2 - s = \frac{t}{2}\left(1 - \sqrt{\frac{t - 4m^2}{t}} \cos\theta\right), \qquad (2.134)$$

$$m^2 - u = \frac{t}{2}\left(1 + \sqrt{\frac{t - 4m^2}{t}} \cos\theta\right). \qquad (2.135)$$

Transposing s and u is equivalent to redefining the angle: $\theta \to \pi - \theta$.

We will discuss the behaviour of the cross section in two special cases: at the threshold, $t \approx 4m^2$, and at very high energies $t \gg 4m^2$.

2.7.1 Annihilation near threshold

In the threshold region, $t \approx 4m^2$, we have $d\Gamma \to \infty$ with $t \to 4m^2$ and the cross section becomes very large ($d\sigma \to \infty$). Physically this can be understood as follows. The region $t - 4m^2 \ll m^2$ corresponds to very slow incident particles whose flux is very small: $j \propto v \ll 1$. The cross section is the ratio of the probability of the process and the flux,

$$d\sigma \propto \frac{|W|^2}{j},$$

and therefore $d\sigma \to \infty$. One can ask why have we not observed a situation like this in the case of elastic scattering, where the flux also goes to zero at the threshold, $j \to 0$. The reason is that, for elastic scattering, the number of final states is also small when $j \to 0$ so that $|W|^2 \to 0$ and the ratio is finite. In the case of annihilation, final photons always have finite energies, $k_f \geq m$, even in the vicinity of the threshold $t = 4m^2$. The phase volume of these final state photons therefore remains finite (does not vanish) and hence the cross section turns out to be singular at the threshold.

2.7.2 e^+e^- annihilation at very high energies

Now let the annihilation energy be very large: $t \gg 4m^2$. In the case of large $-s \sim -u \sim t/2$ (which corresponds to scattering at large angles $\theta \sim 90°$) the cross section is

$$d\sigma \simeq \frac{e^4}{4t} \frac{d\Omega}{(2\pi)^2}, \qquad (2.136)$$

and decreases with energy.

Consider the situation where one of the momentum transfers is kept finite, for example, $-u = \mathcal{O}(m^2)$ while $-s \simeq t \gg 4m^2$. The cross section in this kinematical region is

$$d\sigma \simeq \frac{e^4}{8} \frac{1}{m^2 - u} \frac{d\Omega}{(2\pi)^2}, \qquad (2.137)$$

and does not depend on energy.

The contribution (2.137) is due to the second diagram, which describes transmutation of an electron into a photon with a finite momentum transfer u. Similarly, when $-s = \mathcal{O}(m^2)$, $-u \simeq t \gg 4m^2$ the main contribution comes from the first graph, and

$$d\sigma \simeq \frac{e^4}{8} \frac{1}{m^2 - s} \frac{d\Omega}{(2\pi)^2}. \qquad (2.138)$$

According to (2.134), (2.135) these regions correspond to very small angles, $\theta = \mathcal{O}(2m/\sqrt{t}) \ll 1$ (finite s) or $\pi - \theta = \mathcal{O}(2m/\sqrt{t})$ (finite u). The situation is similar to that in the case of elastic scattering: the photons produced closely follow the direction of the colliding particles.

Comparing (2.137) and (2.124) we learn another lesson of what happens in relativistic theory: at high energies (and the same momentum transfer) the cross sections of two entirely different processes, Compton scattering and electron–positron annihilation, coincide:

$$d\sigma^{\text{annih.}}(t = E_c^2 \gg m^2) \simeq \frac{e^4}{8(m^2 - u)} \frac{d\Omega}{(2\pi)^2} \simeq d\sigma^{\text{Compt.}}(s = E_c^2 \gg m^2)$$

(the corresponding kinematical regions are marked on the Mandelstam plane below).

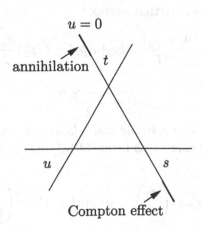

2.8 Electron scattering in an external field

Consider the scattering of an electron off a heavy particle of mass M (a proton, for example). If $m \ll M$ and the momentum transfer q is not very large, this heavy particle will experience almost no recoil. Momenta of the individual particles are related by

$$q = p_1 - p_1' = p_2' - p_2 . \qquad (2.139)$$

If the heavy particle M is initially at rest,

$$p_{20}' \simeq M + \frac{\mathbf{p}_2'^2}{2M} = M + \frac{|\mathbf{q}|^2}{2M}.$$

Its energy practically does not change provided $\mathbf{q}^2/2M \ll M$, so that we can ignore the heavy particle recoil and treat the process as the scattering of the electron by a static external field ($q_0 \simeq 0$).

Let us assume that, like the electron, the heavy particle has spin $\frac{1}{2}$. First, consider photon emission by the heavy particle. Using (2.139) and the Dirac equation, it is easy to prove the identity

$$
\begin{aligned}
2\bar{u}(p_2')\left(\hat{p}_{2\mu} + \hat{p}_{2\mu}'\right)u(p_2) &= \bar{u}(p_2')[(\hat{p}_2 + \hat{p}_2')\gamma_\mu + \gamma_\mu(\hat{p}_2 + \hat{p}_2')]u(p_2) \\
&= \bar{u}(p_2')[(-\hat{q} + 2\hat{p}_2')\gamma_\mu + \gamma_\mu(2\hat{p}_2 + \hat{q})]u(p_2) \\
&= 4M\,\bar{u}(p_2')\gamma_\mu u(p_2) + \bar{u}(p_2')[\gamma_\mu \hat{q} - \hat{q}\gamma_\mu]u(p_2).
\end{aligned}
$$

This gives for the heavy particle vertex

$$\bar{u}(p_2')\gamma_\mu u(p_2) = \frac{(p_2 + p_2')_\mu}{2M}\bar{u}(p_2')u(p_2) - \bar{u}(p_2')\frac{\sigma_{\mu\nu}q_\nu}{2M}u(p_2), \qquad (2.140)$$

where

$$\sigma_{\mu\nu} = \frac{\gamma_\mu\gamma_\nu - \gamma_\nu\gamma_\mu}{2}. \qquad (2.141)$$

We shall calculate the vertex in the non-relativistic limit $M \to \infty$, keeping track of the first order correction terms $\mathcal{O}(|\mathbf{q}|/M) \ll 1$. The Dirac spinors have the form

$$u^\lambda(p_2) = \sqrt{2M}\begin{pmatrix} \varphi_\lambda \\ 0 \end{pmatrix}, \quad u^{\lambda'}(p_2') \simeq \sqrt{2M}\begin{pmatrix} \varphi_{\lambda'} \\ \frac{\boldsymbol{\sigma}\cdot\mathbf{q}}{2M}\varphi_{\lambda'} \end{pmatrix},$$

where we have put $\mathbf{p}_2 = 0$, $\mathbf{p}_2' = \mathbf{q}$, and neglected a quadratic correction $\mathcal{O}(|\mathbf{q}|^2/M^2)$.

Calculating the first term on the right-hand side of (2.140) the lower component of the spinor $u(p_2')$ can be ignored:

$$\bar{u}(p_2')u(p_2) = 2M + \mathcal{O}\left(\frac{\mathbf{p}^2}{M}\right) \simeq 2M. \qquad (2.142)$$

Let us examine

$$\bar{u}(p_2')\,\sigma_{\mu\nu}\,u(p_2).$$

The terms containing $\sigma_{0i} = \gamma_0\gamma_i$ mix the upper and lower components of Dirac spinors and give a small contribution because

$$u^+(p_2')\,\gamma_i\,u(p_2) \;\sim\; 2M\left(\varphi^*, \varphi^*\frac{\boldsymbol{\sigma}\cdot\mathbf{q}}{2M}\right)\begin{pmatrix} 0 \\ -\sigma_i\varphi \end{pmatrix} \;\sim\; M\frac{|\mathbf{q}|}{M},$$

which expression is multiplied by another small factor $\mathcal{O}\,(|\mathbf{q}|/M)$ in (2.140).

We are left with the terms containing matrices with the spatial indices

$$\sigma_{ij} = \gamma_i\gamma_j = \begin{pmatrix} \sigma_i\sigma_j & 0 \\ 0 & \sigma_i\sigma_j \end{pmatrix}$$

sandwiched between the spinors. From the explicit form of the spinors it is clear that the contribution of lower components to the products $\bar{u}\sigma_{ij}u$ is again negligible. Hence, we obtain from (2.140)

$$\bar{u}^{\lambda'}(p_2')\,\gamma_0\,u^\lambda(p_2) = 2M\,\delta_{\lambda,\lambda'},$$
$$\bar{u}^{\lambda'}(p_2')\,\gamma_i\,u^\lambda(p_2) = q_i\,\delta_{\lambda,\lambda'} - q_j\,(\varphi_{\lambda'}^*\,\sigma_i\sigma_j\,\varphi_\lambda),$$

with accuracy up to the linear terms in $|\mathbf{q}|/M$.

Thus, the scattering amplitude of the electron in an external field has the form

$$T = e\,(\bar{u}(p_1')\,\gamma_0\,u(p_1)\cdot A_0(q) - \bar{u}(p_1')\,\gamma_i\,u(p_1)\cdot A_i(q)), \qquad (2.143)$$

where

$$A_0 = e\,\frac{2M}{q^2}, \qquad A_i = e\,\frac{2M}{q^2}\,\varphi^+\left(\frac{q_i}{2M} + i\frac{[\boldsymbol{\sigma}\mathbf{q}]_i}{2M}\right)\varphi. \qquad (2.144)$$

In the derivation of the expression for A_i we have used the identities for the Pauli matrices

$$\sigma_i\sigma_j\,q_j = i\varepsilon_{ijk}\sigma_k\,q_j = -i[\boldsymbol{\sigma}\mathbf{q}]_i.$$

The function $A_0(q)$ is the Fourier component of the Coulomb field generated by the heavy particle.* The first term in the expression for A_i is the Fourier component of the vector potential created by the particle convection current, while the second term is the Fourier component of the vector potential created by its magnetic moment. The term proportional to $\langle e\boldsymbol{\sigma}/2M \rangle$ corresponds to the usual Bohr magneton. In the case of the electron, the magnetic moment coincides with the Bohr magneton.

For the proton one also has to account for the *anomalous* magnetic moment μ_{anom}. Accordingly, we have to write $(1 + \mu_{\text{anom}})[\boldsymbol{\sigma}\mathbf{q}]$ in (2.144) instead of $[\boldsymbol{\sigma}\mathbf{q}]$.

Scattering by an external field corresponds to the limit $M \to \infty$. The expression for the cross section contains $(2M)^2$ in the numerator (coming from $|T|^2$) and $(2M)^2$ in the denominator (coming from the phase volume and the current) which cancel each other. The terms in the amplitude corresponding to the current and the magnetic moment tend to zero for an infinitely heavy target. Hence, the cross section in this case will be determined by the Coulomb potential of the source

$$A_0 = -\frac{e}{\mathbf{q}^2},$$

thus restoring the common normalization of the potential.

2.9 Electron bremsstrahlung in an external field

Due to conservation laws, a free electron cannot emit a photon. The presence of an external field makes such a process possible. Consider the diagrams

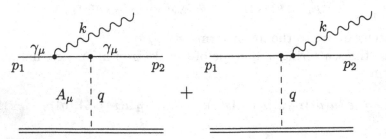

They describe the emission of a photon by an electron before and after the scattering, respectively. The amplitude F_{brems} which corresponds to

---

* Compared with the ordinary scalar and vector potentials, $A_0(q)$ and $A_i(q)$ contain an extra factor $2M$. As we will see in the next paragraph, this extra factor disappears in the calculation of the cross section for electron scattering by a heavy Coulomb source.

these graphs can be written as

$$F_{\text{brems}} = e\,\bar{u}(p_2)\hat{e}\frac{m + \hat{p}_2 + \hat{k}}{m^2 - (p_2 + k)^2}\hat{A}(q)u(p_1)$$

$$+ e\,\bar{u}(p_2)\hat{A}(q)\frac{m + \hat{p}_1 - \hat{k}}{m^2 - (p_1 - k)^2}\hat{e}u(p_1). \tag{2.145}$$

We will consider only the two most interesting cases.

2.9.1 Emission of a soft photon by a low energy electron

In this case we have $p_{10} \simeq m$, $k \ll m$. Due to the pole in the electron Green function $1/(m - \hat{p}_1 + \hat{k})$ the amplitude of this process is very large. Physically this is due to the fact that the emission of soft photons begins far away from the scatterer and takes place over a large region.

Let us calculate the numerators in (2.145) in the soft photon approximation. For the first numerator,

$$\hat{e}(m + \hat{p}_2) + \hat{e}\hat{k} = (m - \hat{p}_2)\hat{e} + 2(ep_2) + \hat{e}\hat{k} \simeq 2(ep_2),$$

since

$$\hat{e}\hat{p} = e_\mu p_\nu \gamma_\mu \gamma_\nu = e_\mu p_\nu \left(-\gamma_\nu \gamma_\mu + 2g_{\mu\nu}\right) = -\hat{p}\hat{e} + 2(ep).$$

Similarly, for the second numerator

$$(m + \hat{p}_1)\hat{e} - \hat{k}\hat{e} \simeq 2(ep_1).$$

Substituting these expressions in (2.145), we obtain

$$F_{\text{brems}} = e\left[\frac{2(ep_2)}{m^2 - (p_2 + k)^2} + \frac{2(ep_1)}{m^2 - (p_1 - k)^2}\right]\bar{u}(p_2)\hat{A}(q)u(p_1).$$

The function

$$f_s(q) = \bar{u}(p_2)\hat{A}(q)u(p_1)$$

is nothing but the electron scattering amplitude in the external field. The bremsstrahlung amplitude becomes

$$F_{\text{brems}} = ef_s(q)\left[\frac{ep_1}{p_1 k} - \frac{ep_2}{p_2 k}\right]. \tag{2.146}$$

In the non-relativistic case $|\mathbf{p}_1| \ll m$, $p_1 k \simeq mk_0$. This gives

$$F_{\text{brems}} = f_s(q)\frac{e}{k_0}\,\mathbf{e}\cdot(\mathbf{v}_2 - \mathbf{v}_1), \tag{2.147}$$

where $\mathbf{v}_{1,2} = \mathbf{p}_{1,2}/m$ are the velocities of the electron before and after scattering. The expression (2.147) coincides with the result given by classical electrodynamics for the bremsstrahlung.

Let us calculate the corresponding cross section:

$$d\sigma_{\text{brems}} = \frac{1}{\mathcal{J}}|F_{\text{brems}}|^2 d^4 p_2 \delta(p_2^2 - m^2)(2\pi)^4 \delta(p_2 + k - p_1 - q)\frac{d^4 k}{(2\pi)^6}\delta(k^2).$$

The cross section for electron scattering in an external field has the form

$$d\sigma_s = \frac{1}{\mathcal{J}}|f_s|^2 \frac{d^4 p_2}{(2\pi)^3}\delta(p_2^2 - m^2)(2\pi)^4 \delta(p_2 + k - p_1 - q).$$

Hence, we may write the bremsstrahlung cross section as

$$d\sigma_{\text{brems}} = d\sigma_s \frac{e^2}{k_0^2}|\mathbf{e}\cdot(\mathbf{v}_2 - \mathbf{v}_1)|^2 \frac{d^4 k}{(2\pi)^3}\delta(k^2).$$

With the help of the relation

$$\frac{d^4 k}{(2\pi)^3}\delta(k^2) = \frac{dk_0 k\, dk^2}{(2\pi)^3 2}\delta(k_0^2 - k^2)d\Omega = \frac{dk_0}{(2\pi)^3}\frac{k_0}{2}d\Omega$$

we get

$$d\sigma_{\text{brems}} = d\sigma_s \frac{e^2}{16\pi^3}|\mathbf{e}\cdot(\mathbf{v}_2 - \mathbf{v}_1)|^2 d\Omega \frac{dk_0}{k_0},$$

or, taking into account that $e^2/4\pi = \alpha = 1/137$,

$$d\sigma_{\text{brems}} = d\sigma_s \frac{\alpha}{2\pi}|\mathbf{e}\cdot(\mathbf{v}_2 - \mathbf{v}_1)|^2 \frac{d\Omega}{2\pi}\frac{dk_0}{k_0}. \qquad (2.148)$$

We see that $(d\sigma_{\text{brems}}/dk_0) \to \infty$ when $k_0 \to 0$, and the total cross section is logarithmically divergent in the small frequency region. Integrating the cross section over photon energy from $k_{0\,\text{min}}$ to $k_{0\,\text{max}}$ we obtain

$$d\sigma_t = d\sigma_s \frac{\alpha}{2\pi}|\mathbf{e}\cdot(\mathbf{v}_2 - \mathbf{v}_1)|^2 \cdot 2\ln\frac{k_{0\,\text{max}}}{k_{0\,\text{min}}}. \qquad (2.149)$$

The frequencies of the emitted photons are limited from above by the energy of the electron, $k_{0\,\text{max}} < p_{10} \sim m$. The lower limit, $k_{0\,\text{min}}$, however, can be chosen within our approximation to be arbitrarily small.

An attempt to include very soft photons by putting $k_{0\,\text{min}} = 0$ immediately leads to a difficulty in the form of an infinite cross section. This problem is called the infrared catastrophe.

The reason for it is pretty obvious: we are attempting to use the lowest order approximation in the coupling constant α in the region where it is not valid any more. Indeed, (2.149) establishes the criterion for the applicability of the lowest order approximation in α. It is valid only when the bremsstrahlung cross section may be considered as a *correction* to the cross section of the process without bremsstrahlung, i.e. when

$$\frac{\alpha}{\pi} \ln \frac{k_{0\,max}}{k_{0\,min}} < 1.$$

If this inequality is violated, processes with emission of more bremsstrahlung photons are not suppressed and should be taken into account.

We can easily understand this from another perspective. At $k \to 0$ we have a classical electromagnetic field which means the presence of a large number of photons. Naturally, in this case we cannot restrict ourselves to a process with the emission of only one photon. We shall return to this problem later when we discuss higher order corrections.

2.9.2 Soft radiation off a high energy electron

Now consider the second case of bremsstrahlung when

$$p_{10} \gg m, \quad \frac{k_0}{p_{10}} \ll 1. \tag{2.150}$$

Neglecting the terms $\hat{e}\hat{k}$ in the numerators of (2.145), we get, as before,

$$F_{brems} = e f_s(q) \left[\frac{e p_1}{p_1 k} - \frac{e p_2}{p_2 k} \right].$$

Further, for the photon emitted at angle θ_1, we have

$$(p_1 k) = p_{10} k_0 - |\mathbf{p}_1| k_0 \cos\theta_1 = |\mathbf{p}_1| k_0 \left[1 - \cos\theta_1 + \frac{m^2}{2\mathbf{p}_1^2} \right],$$

since, due to (2.150), p_{10} can be expanded as

$$p_{10} = \sqrt{m^2 + \mathbf{p}_1^2} \simeq |\mathbf{p}_1| + \frac{m^2}{2|\mathbf{p}_1|}.$$

For small emission angles $1 - \cos\theta_1 \simeq \theta_1^2/2$, and introducing $\theta_0^2 = m^2/\mathbf{p}_1^2$ we can write

$$(p_1 k) \simeq \frac{|\mathbf{p}_1| k_0}{2} \left(\theta_1^2 + \theta_0^2 \right). \tag{2.151}$$

If $\theta_1 \ll 1$, $\theta_0 \ll 1$, the denominator in the amplitude will again be small and the probability of bremsstrahlung will be large. In other words, the bremsstrahlung photons are emitted mainly at small angles.

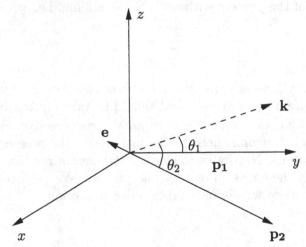

The numerator in the expression for the amplitude equals

$$(ep_1) = -|\mathbf{p}_1|\sin\theta_1 \simeq -|\mathbf{p}_1|\theta_1$$

(the photon polarization lies in the scattering plane $\{\mathbf{p}_1, \mathbf{p}_2\}$). Similar expressions can be obtained for the second term of the amplitude. The result is

$$F_{\text{brems}} = -f_s(q)\,\frac{2e}{k}\left[\frac{\theta_1}{\theta_1^2 + \theta_0^2} - \frac{\theta_2}{\theta_2^2 + \theta_0^2}\right]. \qquad (2.152)$$

The large electron momenta have cancelled.

The expression (2.152) shows that if the electron is scattered at a small angle, $\theta_s \ll \theta_1 \simeq \theta_2$, the photon emission amplitudes before and after scattering are subtracted from each other, and there is practically no bremsstrahlung. (No scattering – no radiation.) If, however, the scattering goes at a sufficiently large angle, $\theta_s \gg \theta_0$, one of the two amplitudes can be much larger than the other, so that the cancellation will no longer occur. This happens in two cases: $\theta_2 \simeq \theta_s \gg \theta_1$ or, vice versa, $\theta_1 \simeq \theta_s \gg \theta_2$.

This means that in the relativistic case the bremsstrahlung is concentrated inside two narrow cones, $\theta_1 \ll \theta_s$, $\theta_2 \ll \theta_s$, the axes of which are directed along $\mathbf{p}_1$ and $\mathbf{p}_2$. These cones are absolutely identical.

As an example, let us consider the photon emission cross section in one of these cones:

$$d\sigma_{\text{brems}} = d\sigma_s\,\frac{4e^2}{k^2}\,\frac{\theta_1^2}{(\theta_1^2 + \theta_0^2)^2}\,\frac{d^4k\,\delta(k^2)}{(2\pi)^3},$$

or, using the relationship

$$\frac{d^4k\,\delta(k^2)}{(2\pi)^3} = \frac{k_0\,dk_0}{(2\pi)^3 2}\,2\pi d\cos\theta,$$

$$d\sigma_{\text{brems}} = d\sigma_s \frac{2e^2}{8\pi^3} \frac{dk_0}{k_0} 2\pi \sin\theta_1 d\theta_1 \frac{\theta_1^2}{(\theta_1^2 + \theta_0^2)^2} \simeq d\sigma_s \frac{\alpha}{\pi} \frac{dk_0}{k_0} \frac{\theta_1^2 d\theta_1^2}{(\theta_1^2 + \theta_0^2)^2}.$$

In the case of large-angle electron scattering, $\theta_s \sim 1$, integrating over the photon energies and photon angles in the cone $\theta_0 < \theta_1 < \theta_s$, we get the total bremsstrahlung cross section in the form

$$d\sigma_t = d\sigma_s \frac{\alpha}{\pi} \ln \frac{k_{0\,\text{max}}}{k_{0\,\text{min}}} \ln \frac{\theta_s^2}{\theta_0^2} \simeq d\sigma_s \frac{\alpha}{\pi} \ln \frac{|\mathbf{p}_1|}{k_{0\,\text{min}}} \ln \frac{\mathbf{p}_1^2}{m^2}, \qquad (2.153)$$

since $k_{0\,\text{max}} \sim |\mathbf{p}_1|$, $\theta_0 = m^2/2\mathbf{p}_1^2$.

Hence, the total cross section of photon bremsstrahlung accompanying large-angle electron scattering grows with the energy of the projectile as the logarithm squared. This invalidates our single-photon approximation at large energies as well.

2.10 The Weizsäcker–Williams formula

Consider the following situation. Let a light particle hit a heavy one (for example, a nucleus or a proton). In the course of the scattering process, various particles (systems of particles) may be created (for example, in the previous section we considered emission of a photon):

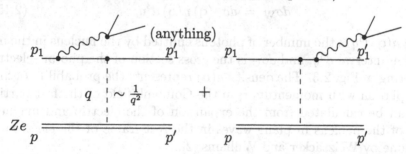

Fig. 2.2

The amplitude for Coulomb scattering of a particle contains the factor $1/q^2$ $(q^2 = (p-p')^2)$, i.e. the main contribution to the cross section comes from small q^2. Can we not derive some general conclusions about such arbitrary processes if the energy of the incoming particle is large, and the momentum transfer q^2 is small?

Suppose that

$$\frac{|q^2|}{m^2} \ll 1 \qquad \text{and} \qquad \frac{s}{m^2} \gg 1.$$

For small q^2, the photon is almost real. Then the scattering can be considered as a two-stage process: first the nucleus emits a photon, then a particle is scattered by this photon, and arbitrary particles are emitted, i.e.

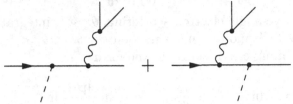

Fig. 2.3

Let us consider the whole process in the rest frame of the electron $(p_1 = (m, 0))$. In this reference frame the nucleus with a high velocity hits the electron.

The Coulomb field of a fast particle is compressed in the direction of motion. We will show that this field can be represented by an ensemble of almost real photons. In this case the cross section for the whole process in Fig. 2.2 may be written as

$$d\sigma_{\mathrm{W}} = d\sigma_{\mathrm{C}}(\mathbf{q})\, n(\mathbf{q})\, d^3q, \qquad (2.154)$$

where $n(\mathbf{q})d^3q$ is the number of photons emitted by the nucleus in the momentum interval d^3q, and $d\sigma_{\mathrm{C}}$ is the cross section of the photon–electron scattering in Fig. 2.3. The density $n(\mathbf{q})$ represents the probability of finding a photon with momentum $\mathbf{q}$ in the Coulomb field of the fast particle and can be calculated from the expansion of the electric and magnetic fields of the nucleus in plane waves in the rest frame of the electron, as was done by Weizsäcker and Williams [2].

We will calculate this density in a different way. The amplitude for the real photon $(q^2 = 0)$ scattering off an electron in Fig. 2.3 is

$$F_{\mathrm{C}} = M_\mu(q, p_1, \ldots)\, e_\mu, \qquad (2.155)$$

where $q^2 = 0$, since the photon is real. On the other hand the electron–proton scattering amplitude in Fig. 2.2 can be written as

$$F_{\mathrm{W}} = \frac{Ze}{q^2}(p + p')_\mu M_\mu(q, p_1, \ldots). \qquad (2.156)$$

Let us find the connection between these amplitudes. If one could assume that in (2.156) q^2 equals zero everywhere except the pole factor, the factors M_μ in these two expressions would coincide. This is a reasonable assumption, since $|q^2|/m^2 \ll 1$ and so q^2 is small compared to all other momenta entering M_μ. Due to current conservation we have

$$q_\mu M_\mu = 0\,, \qquad e_\mu q_\mu = 0. \tag{2.157}$$

Let us choose the z-axis in the rest frame of the electron along the direction of the proton momentum:

$$p = (p_0, p_z, 0, 0), \tag{2.158}$$

and

$$p - q = p'\,, \qquad p^2 = p'^2 = M^2.$$

Then,

$$(p - q)^2 = p^2 = M^2 \qquad \Longrightarrow \qquad -2pq + q^2 = 0$$

or, due to (2.158),

$$-2(p_0 q_0 - p_z q_z) + q^2 = -2p_z(q_0 - q_z) - \frac{2M^2}{p_0 + p_z} q_0 + q^2 = 0. \tag{2.159}$$

The proton momentum is large, $p_0 \simeq p_z \gg M$, and we derive from (2.159)

$$q_0 - q_z \simeq \frac{q^2}{2p_0} - \frac{M^2 q_0}{2p_0^2}. \tag{2.160}$$

This difference is very small, $|q_0 - q_z| \propto p_0^{-1}$, while the photon momentum components may be rather large, $q_0, q_z \gg m$ (up to $q_0, q_z \propto p_0$).

Let us call $\mathbf{q}_\perp$ the component of the photon momentum in the plane perpendicular to the z-axis. Then we have ($q_\perp \equiv |\mathbf{q}_\perp|$)

$$q^2 = q_0^2 - q_z^2 - q_\perp^2 \simeq 2q_0(q_0 - q_z) - q_\perp^2, \tag{2.161}$$

which shows that $q^2 \simeq -q_\perp^2$. This means that the virtual photon is relativistic: $q_0 \simeq q_z \gg m \gg q_\perp \simeq \sqrt{-q^2}$.

Now consider the factor

$$(p + p')_\mu M_\mu(q, p_1, \dots) = (2p - q)_\mu M_\mu$$

in (2.156). Due to current conservation, the right-hand side equals

$$2p_\mu M_\mu \simeq 2p_0(M_0 - M_z).$$

At the same time,

$$q_\mu M_\mu = q_0 M_0 - q_z M_z - \mathbf{q}_\perp \cdot \mathbf{M}_\perp \simeq q_0(M_0 - M_z) - \mathbf{q}_\perp \cdot \mathbf{M}_\perp = 0,$$

or

$$M_0 - M_z = \frac{\mathbf{q}_\perp \cdot \mathbf{M}_\perp}{q_0}.$$

For the amplitude in (2.156) we thus obtain

$$(p + p')_\mu M_\mu = \frac{2p_0}{q_0}\, \mathbf{q}_\perp \cdot \mathbf{M}_\perp. \qquad (2.162)$$

Hence, the amplitude for electron–proton scattering depends only on the transverse projection of the virtual photon–proton amplitude. Physically this means that what makes our virtual exchange photons quasi-real is not only small q^2 but also that their polarizations are practically the same as for the real photons.

Now consider scattering amplitude $e_\mu M_\mu$ in (2.155) for the real photon. In the radiation (Coulomb) gauge

$$\mathbf{e} \cdot \mathbf{q} = 0 \qquad \text{and} \qquad e_0 = 0. \qquad (2.163)$$

This gives

$$e_\mu M_\mu = -e_z M_z - \mathbf{e}_\perp \cdot \mathbf{M}_\perp.$$

However,

$$e_z q_z + \mathbf{e}_\perp \cdot \mathbf{q}_\perp = 0,$$

and due to the smallness of the ratio $q_\perp/q_z \ll 1$, the longitudinal component e_z of the polarization vector is also small:

$$e_z = -\frac{\mathbf{e}_\perp \cdot \mathbf{q}_\perp}{q_z} \ll e_\perp.$$

This means that in our kinematics the real photon polarization vectors are transverse not only with respect to the photon momentum, but they are practically orthogonal to the momentum of the fast particle as well. Hence, the scattering amplitude in (2.155) depends only on $\mathbf{e}_\perp$, and we have

$$e_\mu M_\mu \simeq -\mathbf{e}_\perp \cdot \mathbf{M}_\perp. \qquad (2.164)$$

Thus, physics is completely determined by the transverse part $\mathbf{M}_\perp$ of the photon scattering amplitude both for the virtual photon in (2.162) and for the real photon in (2.155). Note that $q_\perp \ll m$, and the amplitude $\mathbf{M}_\perp$ does not depend on the *direction* of $\mathbf{q}_\perp$. Calculating the cross section we are going to average over two transverse polarizations $\mathbf{e}_\perp$ ($\mathbf{e}_\perp^2 = 1$) for the real photons, and integrate over all transverse directions for the virtual ones. Bearing this in mind, we can simply use the normalized vectors $\mathbf{q}_\perp/\sqrt{q_\perp^2}$ as the polarization vectors in (2.155), and the amplitude (2.156) may be written as

$$(p + p')_\mu M_\mu = \frac{2p_0}{q_0}\, q_\perp \left(\mathbf{e}_\perp \cdot \mathbf{M}_\perp\right) = -\frac{2p_0}{q_0}\, q_\perp F_{\mathrm{C}},$$

and

$$F_W = -\frac{Ze}{q^2}\frac{2p_0}{q_0}q_\perp F_C.$$ (2.165)

The cross section $d\sigma_W$ for electron–proton scattering in this approximation has the form

$$d\sigma_W = \frac{Z^2e^2}{q^4}\left(\frac{2p_0}{q_0}\right)^2 q_\perp^2 \frac{1}{4mp_0}\frac{d^4p'}{(2\pi)^3}\delta_+(p'^2 - M^2) \times 4mq_0$$

$$\times \left[\frac{1}{4mq_0}|F_C|^2 \frac{d^4k_1\delta_+(k_1^2 - m_1^2)\dots d^4k_n\delta_+(k_n^2 - m_n^2)}{(2\pi)^{3n}}\right.$$ (2.166)

$$\left. \times (2\pi)^4\delta\left(p_1 + p - \sum k_i - p' - p_1'\right)\frac{d^4p_1'}{(2\pi)^3}\delta(p_1'^2 - m^2)\right].$$

Taking into account $p - p' = q$, we have

$$\delta\left(p_1 + p - \sum k_i - p' - p_1'\right) = \delta\left(p_1 + q - \sum k_i - p_1'\right).$$

Observing that the expression in the square brackets in (2.166) is nothing but the cross section $d\sigma_C$ of the electron–photon interaction in Fig. 2.3 we arrive at

$$d\sigma_W = \frac{Z^2e^2}{q^4}\left(\frac{2p_0}{q_0}\right)^2 q_\perp^2 \frac{q_0}{p_0} d\sigma_C(q)\frac{d^4q}{(2\pi)^3}\delta(-2pq + q^2),$$

$$d^4q = dq_0dq_zd^2q_\perp.$$

Integration over q_0 is trivial due to the δ-function (which gives $2p_0$ in the denominator), and we obtain

$$d\sigma_W = n(q)\, d\sigma_C\, dq_z d^2q_\perp,$$ (2.167)

where

$$n(q) = \frac{Z^2e^2}{(2\pi)^3}\frac{2q_\perp^2}{q_0q^4} = \frac{Z^2\alpha}{\pi^2}\frac{q_\perp^2}{q_0q^4}$$ (2.168)

is the momentum-space density of photons emitted by the proton. Let us write the cross section (2.167) (using $q_0 \simeq q_z$) as

$$d\sigma_W = d\sigma_C \frac{Z^2\alpha}{\pi^2}\frac{dq_0}{q_0}\frac{q_\perp^2 d^2q_\perp}{q^4}.$$ (2.169)

We started the discussion of electron–proton scattering with the idea that the region of small photon virtualities $|q^2| \ll m^2$ (small momentum transfer $q_\perp \ll m$) gives a large contribution to the cross section. Let us check if this is what we have obtained.

Combining (2.160) and (2.161) we see that

$$-q^2\left(1-\frac{q_0}{p_0}\right)\simeq q_\perp^2+\frac{M^2q_0^2}{p_0^2}.$$

Then $q^4\simeq q_\perp^4$ in a wide interval of momenta

$$\frac{q_0}{p_0}\ll\frac{m}{M}<1,\qquad\frac{q_0^2M^2}{p_0^2}\ll q_\perp^2\ll m^2,$$

and the integral over $q_\perp$ in (2.169) is logarithmically enhanced:

$$d\sigma_{\rm W}=d\sigma_{\rm C}\,\frac{Z^2\alpha}{\pi}\frac{dq_0}{q_0}\ln\frac{p_0^2m^2}{q_0^2M^2}.\qquad(2.170)$$

We still have freedom to transfer different energies. Integrating our logarithmic distribution (2.170) over q_0 over a wide interval of energies $m\ll q_0\ll p_0m/M$, we obtain[†]

$$d\sigma_{\rm W}=d\sigma_{\rm C}\frac{Z^2\alpha}{\pi}\ln^2\frac{p_0^2}{M^2}.\qquad(2.171)$$

The double-logarithmic enhancement factor makes this cross section large at large energies. The enhancement is due to a large number of quasi-real photons surrounding a fast proton, and the large density of these photons compensates for possible smallness of the cross section per photon. Hence, scattering of a fast particle with small momentum transfer (the Weizsäcker–Williams-type process) can serve as an intense source for production of different particles at large energies.

The very first experimental lower limit for the mass of the W-boson was derived long ago just from a WW-type process. The weak interaction looks as follows:

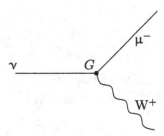

How can the W-boson be detected? The neutrino practically does not interact. However, due to the electromagnetic interaction of the muon

---

[†] Here the cross section of the Compton-type process has a finite high-energy limit $d\sigma_{\rm C}(s)\to$ const for $s=(p'+q)^2\simeq 2mq_0\gg m^2$, see Section 2.6.2.

with the nucleus

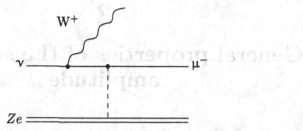

the cross section for scattering on the Coulomb field of the nucleus is large. From this process the bound $m_W > 5$ GeV has been obtained.

3

General properties of the scattering amplitude

3.1 Symmetries in quantum electrodynamics

Quantum electrodynamics is relativistically invariant, it was constructed this way. In addition, QED is invariant with respect to some discrete transformations which cannot be reduced to the Lorentz transformations proper. They are

P – inversion of the space coordinates: $\mathbf{x}' = -\mathbf{x}$,

T – the time reversal: $t' = -t$, and

C – charge conjugation, i.e. replacing all particles by antiparticles. (Existence of antiparticles does not imply that they interact in the same way as particles. We know that *free* particles and antiparticles are indistinguishable, but the similarity may end there.)

We shall show that QED is invariant under each of these three symmetry operations.

3.1.1 P-conservation

Under reflection of space coordinates, momentum p_μ of the electron transforms as

$$(p_0, \mathbf{p}) \xrightarrow{P} (p_0', \mathbf{p}') = (p_0, -\mathbf{p}),$$

since $\mathbf{p}$ is an ordinary three-dimensional vector, while the energy p_0 does not change, because it depends only on velocity squared $\mathbf{v}^2$.

The electron is described not only by its momentum but also by its spin ζ_μ; $p_\mu \zeta_\mu = 0$, $\zeta^2 = -1$. For the electron at rest ζ_μ has only spatial components,

$$\zeta_\mu = (0, \boldsymbol{\zeta}).$$

For the electron moving with velocity **v** the spin four-vector becomes

$$\zeta'_0 = \frac{(\mathbf{v} \cdot \boldsymbol{\zeta})}{\sqrt{1 - \mathbf{v}^2}},$$

$$\zeta'_{\parallel} = \frac{\zeta_{\parallel}}{\sqrt{1 - \mathbf{v}^2}}, \quad \zeta'_{\perp} = \zeta_{\perp}. \tag{3.1}$$

How does the electron spin transform under the P-inversion? Due to similarity between spin and the classical angular momentum, $\mathbf{J} = [\mathbf{r} \times \mathbf{p}]$, we conclude that $\boldsymbol{\zeta}$ is a *pseudovector*, that is, it does not change sign under the spatial inversion. From (3.1) it is then clear that ζ_0 changes sign together with velocity. So we have

$$p_\mu \xrightarrow{P} p'_\mu = (p_0, -\mathbf{p}),$$

$$\zeta_\mu \xrightarrow{P} \zeta'_\mu = (-\zeta_0, \boldsymbol{\zeta}). \tag{3.2}$$

Now we can compare the amplitudes of one and the same process before and after the spatial inversion (that is, the process in two coordinate systems with opposite senses). Consider the scattering process

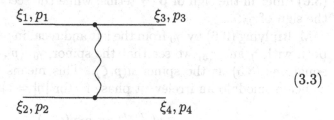

$$\xi_1, p_1 \qquad\qquad \xi_3, p_3$$

$$\xi_2, p_2 \qquad\qquad \xi_4, p_4 \tag{3.3}$$

which after inversion of the spatial coordinates turns into

$$\xi'_1, p'_1 \qquad\qquad \xi'_3, p'_3$$

$$\xi'_2, p'_2 \qquad\qquad \xi'_4, p'_4 \tag{3.4}$$

where the dashed momentum and spin variables in (3.4) are connected with those in (3.3) according to (3.2). Symmetry with respect to space reflection would mean equality of these amplitudes.

Let us check if this is the case. The two amplitudes differ only in the spinor factors. The upper line in (3.2) contains the factor

$$\bar{u}(p_3, \zeta_3)\gamma_\mu u(p_1, \zeta_1),$$

while the respective factor in (3.4) is

$$\bar{u}(p_3', \zeta_3')\gamma_\mu u(p_1', \zeta_1').$$

How do these factors differ? The spinors $u(p, \zeta)$ are determined by the equations (2.29):

$$(\hat{p} - m)u(p, \zeta) = 0,$$
$$(\gamma_5 \hat{\zeta} - 1)u(p, \zeta) = 0. \tag{3.5}$$

Similarly, for $u(p', \zeta')$ we have

$$(\hat{p}' - m)u(p', \zeta') = 0,$$
$$(\gamma_5 \hat{\zeta}' - 1)u(p', \zeta') = 0. \tag{3.6}$$

To relate the two spinors we observe that the first equations in (3.5) and (3.6) differ in the sign of $\mathbf{p} \cdot \boldsymbol{\gamma}$ terms, while the second equations differ in the sign of $\gamma_5 \zeta_0$.

Multiplying (3.6) by γ_0 from the left and recalling that γ_0 anticommutes both with $\boldsymbol{\gamma}$ and γ_5, we see that the spinor $\gamma_0 u(p', \zeta')$ satisfies the same equations (3.5) as the spinor $u(p, \zeta)$. This means that the two spinors coincide (modulo an irrelevant phase factor $|\eta| = 1$):

$$u(p', \zeta') = \gamma_0 u(p, \zeta). \tag{3.7}$$

Similarly, for the Dirac conjugate spinors one obtains

$$\bar{u}(p', \zeta') = \bar{u}(p, \zeta)\gamma_0. \tag{3.8}$$

Thus, the vertex function in (3.4) takes the form

$$\Gamma_\mu' = \bar{u}(p_3', \zeta_3')\gamma_\mu u(p_1', \zeta_1') = \bar{u}(p_3, \zeta_3)\,\gamma_0\gamma_\mu\gamma_0\,u(p_1, \zeta_1), \tag{3.9}$$

i.e.

$$\Gamma_0' = \Gamma_0, \quad \Gamma_i' = -\Gamma_i. \tag{3.10}$$

We see that under spatial inversion the electron–photon vertex changes. However, our scattering diagram includes two vertices, and the amplitude describing the scattering process remains unchanged. (The same is true for any diagram with virtual photons which, obviously, has an even number of vertices.)

In the diagrams that include photon(s) in the initial/final state, the corresponding interaction vertex is always multiplied by the polarization vector e_μ^λ of a real photon. In this case

$$\Gamma'_\mu e'^\lambda_\mu = \Gamma_\mu e^\lambda_\mu,$$

since e_μ^λ is an ordinary vector and its space components also change sign under space inversion:

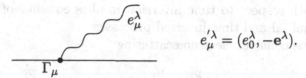

$$e'^\lambda_\mu = (e_0^\lambda, -\mathbf{e}^\lambda).$$

We conclude that electrodynamics as a whole is P-invariant, because it is constructed on the basis of the vector vertex 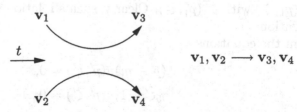 .

It is easy to construct an interaction vertex that violates P-parity. Consider, for example, V–A interaction, described by the vertex

$$\Gamma = \bar{u}\gamma_\mu(1 - \gamma_5)u. \tag{3.11}$$

Repeating the same steps as for the vector vertex above we would obtain the V–A vertex in the reflected reference frame (compare with (3.9)),

$$\Gamma' = \bar{u}\gamma_0\,\gamma_\mu(1 - \gamma_5)\,\gamma_0 u = \bar{u}\,\gamma_0\gamma_\mu\gamma_0\,(1 + \gamma_5)\,u, \tag{3.12}$$

Due to the opposite relative sign between the entries in the bracket $(1+\gamma_5)$ this spatially reflected vertex is essentially different from (3.11). This means that the V–A interaction does not conserve parity, and in this case one can experimentally distinguish *left* from *right* (the right- and left-handed reference frames). This is possible (and not so strange) if a particle lacks the left–right symmetry itself, i.e. has an inherent *chirality* (like a screw).

This is exactly what happens in real *weak interaction* with the basic vertex (3.11).

3.1.2 T-invariance

A classical scattering process

$$\mathbf{v}_1, \mathbf{v}_2 \longrightarrow \mathbf{v}_3, \mathbf{v}_4$$

after time inversion $t \to t' = -t$ turns into the process

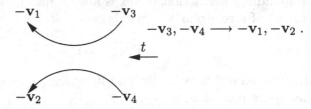

$$-\mathbf{v}_3, -\mathbf{v}_4 \longrightarrow -\mathbf{v}_1, -\mathbf{v}_2 \,.$$

Invariance with respect to time inversion implies equality of the amplitudes of the initial and time-inverted processes.

Again, let us consider electron scattering:

$$ \tag{3.13} $$

Obviously, after time inversion $p' = (p_0, -\mathbf{p})$. To determine what happens with spin variables, we again use the similarity between spin and the classical angular momentum $\mathbf{J} = [\mathbf{r} \times \mathbf{p}]$. Under time inversion the classical angular moment changes sign together with momentum, and so should the quantum spin vector $\boldsymbol{\zeta}$. We have, therefore,

$$
\begin{aligned}
p_\mu &\xrightarrow{P} p'_\mu = (p_0, -\mathbf{p}), \\
\zeta_\mu &\xrightarrow{P} \zeta'_\mu = (\zeta_0, -\boldsymbol{\zeta}).
\end{aligned}
\tag{3.14}
$$

Do the amplitudes (3.13) coincide?

Let us compare the upper vertices in the two diagrams:

$$ \Gamma_\mu = \bar{u}(p_3, \zeta_3)\gamma_\mu u(p_1, \zeta_1) \,, \tag{3.15} $$

and

$$ \Gamma'_\mu = \bar{u}(p'_1, \zeta'_1)\gamma_\mu u(p'_3, \zeta'_3) = u^\top(p'_3, \zeta'_3)\gamma_\mu^\top \bar{u}^\top(p'_1, \zeta'_1) \,, \tag{3.16} $$

where we write the latter vertex in terms of transposed quantities to change the order of spinors. We have to connect $u^\top(p'_3, \zeta'_3)$ with $\bar{u}(p_3, \zeta_3)$, and $\bar{u}(p_1, \zeta_1)$ with $\bar{u}^\top(p'_1, \zeta'_1)$. Clearly, such a relation will contain complex conjugation.

From the equations

$$
\begin{aligned}
(\hat{p}' - m)u(p', \zeta') &= 0, \\
(\gamma_5 \hat{\zeta}' - 1)u(p', \zeta') &= 0,
\end{aligned}
$$

we obtain the equations for the transposed spinors:

$$u^\top(p',\zeta')(\hat{p}'^\top - m) = 0,$$
$$u^\top(p',\zeta')(\hat{\zeta}'^\top\gamma_5^\top - 1) = 0. \tag{3.17}$$

Ordinary equations for the Dirac conjugated spinors are

$$\bar{u}(p,\zeta)(\hat{p} - m) = 0,$$
$$\bar{u}(p,\zeta)(\gamma_5\hat{\zeta} - 1) = 0. \tag{3.18}$$

The first equations of (3.17) and (3.18) contain

$$p_0\gamma_0^\top + \mathbf{p}\cdot\boldsymbol{\gamma}^\top = p_0\gamma_0 - p_1\gamma_1 + p_2\gamma_2 - p_3\gamma_3 \tag{3.19}$$

and

$$p_0\gamma_0 - \mathbf{p}\cdot\boldsymbol{\gamma} = p_0\gamma_0 - p_1\gamma_1 - p_2\gamma_2 - p_3\gamma_3, \tag{3.20}$$

respectively. In (3.19) we have used $\gamma_1^\top = -\gamma_1$, $\gamma_3^\top = -\gamma_3$, while the matrix γ_2 is invariant under transposition, $\gamma_2^\top = \gamma_2$ (together with $\gamma_0^\top = \gamma_0$).

In the case of P-parity considered above, we managed to 'equalize' the equations (3.5) and (3.6) by multiplying the spinor $u(p',\zeta')$ by γ_0. Now, to obtain a relationship between $\bar{u}(p,\zeta)$ and $u^\top(p',\zeta')$, we need somehow to change the sign of γ_2 in (3.19). This can be achieved by multiplying this equation from the right by the matrix $i\gamma_0\gamma_1\gamma_3$. (We have inserted i here to ensure that nothing changes after the *double* time reflection: $(i\gamma_0\gamma_1\gamma_3)^2 = 1$.)

Indeed, after multiplication by $i\gamma_0\gamma_1\gamma_3$ (3.19) turns into (3.20):

$$(p_0\gamma_0 - p_1\gamma_1 + p_2\gamma_2 - p_3\gamma_3)\gamma_0\gamma_1\gamma_3 = \gamma_0(p_0\gamma_0 + p_1\gamma_1 - p_2\gamma_2 + p_3\gamma_3)\gamma_1\gamma_3$$
$$= \gamma_0\gamma_1(-p_0\gamma_0 + p_1\gamma_1 + p_2\gamma_2 - p_3\gamma_3)\gamma_3 = \gamma_0\gamma_1\gamma_3(p_0\gamma_0 - p_1\gamma_1 - p_2\gamma_2 - p_3\gamma_3).$$

It is straightforward to verify that the same operation 'equalizes' also the equations for spin vectors – the second lines in (3.17) and (3.18).

Thus, $u^\top i\gamma_0\gamma_1\gamma_3$ satisfies (3.18), and we can identify the two spinors:

$$u^\top(p',\zeta')\, i\gamma_0\gamma_1\gamma_3 = \bar{u}(p,\zeta).$$

After simple algebra we arrive at

$$u^\top(p',\zeta') = -i\,\bar{u}(p,\zeta)\,\gamma_3\gamma_1\gamma_0,$$
$$\bar{u}^\top(p',\zeta') = i\,\gamma_0\gamma_1\gamma_3\,u(p,\zeta). \tag{3.21}$$

Inserting (3.21) into the vertex (3.16), we obtain

$$\Gamma'_\mu(p',\zeta') = \bar{u}(p,\zeta)\,\gamma_3\gamma_1\gamma_0\gamma_\mu^\top\gamma_0\gamma_1\gamma_3\,u(p,\zeta), \tag{3.22}$$

which results in

$$\Gamma_0' = \Gamma_0 , \quad \Gamma_i' = -\Gamma_i . \tag{3.23}$$

We see that the time inversion changes the electron–photon vertex the same way the space reflection does. Our scattering diagram, however, includes two vertices, and the amplitude describing the scattering process coincides with the amplitude for the time-inverted one. (Which is true for any diagram with virtual photons only.)

In processes involving real photons, an external vertex is always multiplied by a polarization vector e_μ^λ. Spatial components of the latter change sign under time reversal*, so that

$$\Gamma_\mu' e_\mu'^\lambda = \Gamma_\mu e_\mu^\lambda . \tag{3.24}$$

Hence, electrodynamics is T-invariant.

3.1.3 C-invariance

Is quantum electrodynamics invariant under replacement of particles by antiparticles? To answer this question we need to compare the processes for particles with those for antiparticles with the same momenta but with opposite polarizations. (The polarizations should be opposite since we have already found in Section 2.1 that the transition from particle to antiparticle implies $\zeta' = -\zeta$.)

Let us compare the scattering amplitudes

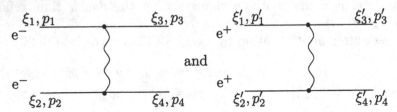

The interaction vertex of the positron scattering amplitude in terms of spinors v has the form

$$\bar{v}(p_1) \gamma_\mu v(p_3), \tag{3.25}$$

where $v(p)$ satisfies

$$(\hat{p} - m)\, v(p) = 0 ,$$

$$\left(-\gamma_5 \hat{\zeta} - 1 \right) v(p) = 0 .$$

---

* Notice that this transformation law respects the condition $(e_\mu k_\mu) = (e_\mu' k_\mu') = 0$ which we imposed on physical photon polarization vectors. In essence, **e** is a vector potential **A** which transforms under time inversion like velocity **v** of the charge it is created by.

Using the relations between v and u spinors following from (2.48),

$$v^\top(p_3) = -\bar{u}(p_3)\,C,$$
$$\bar{v}^\top(p_1) = C^{-1}u(p_1),$$

(3.26)

we derive

$$\Gamma'_\mu = \bar{v}(p_1)\gamma_\mu v(p_3) = v^\top(p_3)\gamma_\mu^\top \bar{v}^\top(p_1) = -\bar{u}(p_3)\,C\gamma_\mu^\top C^{-1}u(p_1) = \Gamma_\mu,$$

(3.27)

since

$$C\gamma_\mu^\top C^{-1} = -\gamma_\mu\,.$$

By reflecting one vertex we turn to the e^+e^- (u-channel) scattering amplitude. In this case, as we have learned in the previous chapter, the *annihilation* amplitude enters with a plus sign, while the *Coulomb scattering* amplitude acquires a minus sign which reflects Fermi statistics of the e^-e^- pair in the s-channel. Thus, although the vertex itself is invariant under charge conjugation, the *amplitude* is C-odd. It is possible to attach this minus sign to the vertex, so that

$$\Gamma_\mu \xrightarrow{C} -\Gamma_\mu\,.$$

(3.28)

This is natural from the point of view of the non-relativistic quantum mechanical analogy, where the photon emission amplitude is proportional to the electric charge of a particle, e for an electron, $(-e)$ for a positron. (It is worthwhile to notice that such a convention reproduces itself in more complicated processes with more than one photon attached to the positron line. Indeed, in adding a photon we add one vertex and one positron propagator to the diagram. According to (3.27), the vertex $\Gamma_{e^+ \to e^+ + \gamma}$, written in terms of the v spinors, is identical to $\Gamma_{e^- \to e^- \gamma}$. The additional minus sign to be ascribed to it then comes from the positron propagator (2.59), so that the prescription (3.28) remains valid.)

Like in the case of other discrete transformations, the fundamental vertex changes after charge conjugation, but our scattering amplitude $A_{e^-e^-} \to A_{e^+e^+}$ does not.

As usual, we have to consider separately the case of an odd number of vertices, that is, the case of external real photons. Since the amplitude (effectively, the vertex) changes sign, (3.28), the theory will be C-invariant if the polarization vector changes sign too:

$$e_\mu \xrightarrow{C} e'_\mu = -e_\mu\,.$$

(3.29)

Discussing P- and T-symmetries, we used a classical analogy between spin and angular momentum to find out how e_μ transforms under these operations. Charge conjugation is a new symmetry that does not have a classical analog. Therefore we simply *postulate* the transformation law (3.29).

This condition can easily be satisfied, since generally speaking the wave function of a neutral particle transforms into itself under charge conjugation only up to a phase factor, which we can always choose at will.

Charge conjugation invariance and negative charge parity of the photon (3.29) impose strong restrictions on electromagnetic processes. For example, the transition of two photons into three is forbidden since the initial two-photon state has positive charge parity, while the wave function of the final state changes sign under charge conjugation.

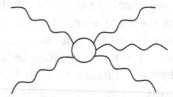

We have obtained the following transformation laws for the wave functions under discrete transformations:

$$P: \quad u(p',\zeta') = \gamma_0 u(p,\zeta),$$
$$T: \quad u^\top(p',\zeta') = -i\,\bar{u}(p,\zeta)\,\gamma_3\gamma_1\gamma_0, \qquad (3.30)$$
$$C: \quad v^\top(p',\zeta') = -i\,\bar{u}(p,\zeta)\,\gamma_2\gamma_0.$$

Quantum electrodynamics is invariant under each of these three transformations.

We have discussed above the weak interaction V–A vertex as an example of a parity violating interaction. Let us consider now how this interaction behaves under time reversal:

$$\gamma_\mu(1-\gamma_5) \implies \gamma_0\gamma_1\gamma_3[\gamma_\mu(1-\gamma_5)]^\top \gamma_3\gamma_1\gamma_0\,.$$

Since $\gamma_5^\top = \gamma_5$ and anticommutes with all γ_μ, it is easy to see that

$$\gamma_0\gamma_1\gamma_3[\gamma_\mu(1-\gamma_5)]^\top \gamma_3\gamma_1\gamma_0 \;=\; \gamma_0\gamma_1\gamma_3\gamma_\mu^\top \gamma_3\gamma_1\gamma_0\,(1-\gamma_5),$$

which means that weak interaction is T-invariant, in the same way as in QED.

At the same time, an interaction of the form, say, $\bar{u}(1+\gamma_5)u$ would violate T-parity:

$$(1+\gamma_5) \overset{T}{\implies} \gamma_0\gamma_1\gamma_3(1+\gamma_5)^\top \gamma_3\gamma_1\gamma_0 = (1-\gamma_5) \neq (1+\gamma_5)\,.$$

Let us return to weak interaction. Under charge conjugation the vertex transforms as

$$C^{-1}[\gamma_\mu(1-\gamma_5)]^\top C = C^{-1}\gamma_\mu^\top(1+\gamma_5)C = -\gamma_\mu(1+\gamma_5). \tag{3.31}$$

We see that weak interaction violates charge conjugation invariance C, as well as P-parity. It respects, however, the so-called *combined inversion*, CP, since each of the two transformations changes the relative sign of γ_5 in (3.12) and (3.31):

$$\gamma_\mu(1-\gamma_5) \overset{P}{\Longrightarrow} \gamma_\mu(1+\gamma_5) \overset{C}{\Longrightarrow} \gamma_\mu(1-\gamma_5).$$

CP-invariance of weak interaction means that although it is P-odd we still cannot distinguish left and right, because we just do not know whether we deal with a particle or an antiparticle. We only know that if what we call a *particle* is left-handed, its *antiparticle* will be right-handed.

3.2 The *CPT* theorem

What will happen if we carry out, one after the other, all three discrete transformations? The momenta and spins stay intact:

$$\begin{aligned}
p &= (p_0, \mathbf{p}) \overset{P}{\longrightarrow} (p_0, -\mathbf{p}) \overset{T}{\longrightarrow} (p_0, \mathbf{p}) \overset{C}{\longrightarrow} (p_0, \mathbf{p}),\\
\zeta &= (\zeta_0, \boldsymbol\zeta) \overset{P}{\longrightarrow} (-\zeta_0, \boldsymbol\zeta) \overset{T}{\longrightarrow} (-\zeta_0, -\boldsymbol\zeta) \overset{C}{\longrightarrow} (\zeta_0, \boldsymbol\zeta).
\end{aligned} \tag{3.32}$$

The spinors under discrete symmetry operations transform according to (3.30), which relations are equivalent to

$$\begin{aligned}
u(p,\zeta) &\overset{P}{\longrightarrow} u(p',-\zeta') = \gamma_0\, u(p,\zeta),\\
u(p,\zeta) &\overset{T}{\longrightarrow} u(p',\zeta') = -i\gamma_1\gamma_3\, u^*(p,\zeta),\\
u(p,\zeta) &\overset{C}{\longrightarrow} v(p,-\zeta) = i\gamma_2\, u^*(p,\zeta)
\end{aligned} \tag{3.33}$$

(where $a' \equiv (a_0, -\mathbf{a})$).

Carrying out the chain of three transformations, we arrive at

$$\begin{aligned}
u(p,\zeta) &\overset{P}{\longrightarrow} u(p',-\zeta') = \gamma_0\, u(p,\zeta)\\
&\overset{TP}{\longrightarrow} u(p,-\zeta) = -i\gamma_1\gamma_3\,(\gamma_0\, u^*(p,\zeta))\\
&\overset{CTP}{\longrightarrow} v(p,\zeta) = i\gamma_2\,(i\gamma_1\gamma_3\gamma_0\, u) = \gamma_1\gamma_2\gamma_3\gamma_0\, u = i\gamma_5\, u(p,\zeta).
\end{aligned} \tag{3.34}$$

Thus, we have obtained

$$u(p,\zeta) \overset{CTP}{\longrightarrow} v(p,\zeta) = i\gamma_5\, u(p,\zeta). \tag{3.35}$$

Diagrammatically the discrete transformations look as follows:

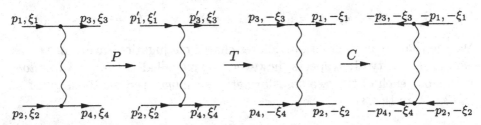

We see that after the CPT transformation all momenta and polarizations in the spinors $u(p, \zeta)$ change signs. Since the particle spinor $u(-p, -\zeta)$ describes the *antiparticle* with momentum p and spin vector ζ, we conclude that the CPT invariance means that the amplitudes of the process

$$p_1, \zeta_1; p_2, \zeta_2 \longrightarrow p_3, \zeta_3; p_4, \zeta_4$$

for particles and antiparticles coincide.

Quantum electrodynamics is obviously CPT invariant, since it is invariant under each of the discrete transformations. CPT invariance of quantum electrodynamics may also be be verified directly using the spinor transformation laws obtained above. Indeed, using the relation (3.35), and its conjugate

$$\bar{v}(p, \zeta) = -i\, \bar{u}(p, \zeta)\, \gamma_5 \, ,$$

we obtain

$$\Gamma'_\mu = \bar{v}(p_1)\gamma_\mu v(p_3) = \bar{u}(p_1)\gamma_5\gamma_\mu\gamma_5 u(p_3) = -\bar{u}(p_1)\gamma_\mu u(p_3).$$

On the other hand,

$$e_\mu = (e_0, \mathbf{e}) \xrightarrow{P} (e_0, -\mathbf{e}) \xrightarrow{T} (e_0, \mathbf{e}) \xrightarrow{C} -e_\mu,$$

and hence

$$\Gamma'_\mu e'_\mu = \Gamma_\mu e_\mu.$$

It is worthwhile to notice that, whatever the structure of the interaction vertex,

$$\Gamma \propto 1, \quad \gamma_5, \quad \gamma_\mu, \quad \gamma_\mu\gamma_5, \quad \gamma_\mu\gamma_\nu \cdots,$$

the law (3.35) guarantees CPT invariance of the interaction.

Is it possible to invent a CPT violating interaction which is, at the same time, relativistically invariant? Obviously not, because the signs of the four-momenta and the polarizations may be changed by a complex Lorentz transformation. In other words, CPT transformation is an element of the Lorentz group if the amplitudes are analytic.

We can thus state that CPT invariance is a fundamental result of our theory, namely, of relativistic invariance and of causality. On the other hand, violation of the invariance under P, T, C, CP, PT, TC does not contradict any fundamental property of the theory, it is just connected with concrete properties of the interacting particles. In fact, none of these symmetries is ever strictly valid. The CPT theorem can be understood in the following way. Particles and antiparticles have built-in screws, clocks and charges. If we associate a definite (right or left) screw, arrow of time and charge with a particle, its antiparticle will have the opposite screw, the opposite arrow of time and the opposite charge. However, all these properties are relative and there is no absolute way to say which object should be called a particle and which an antiparticle. Likewise, there is no way for us to figure out if we are living in World or anti-World.

3.2.1 PT-invariant amplitudes

Let us discuss a useful relation which holds for PT-invariant interactions. PT-transformation does not change the momenta

$$(p_0, \mathbf{p}) \xrightarrow{PT} (p_0, \mathbf{p}),$$

while the spins change signs:

$$(\zeta_0, \boldsymbol{\zeta}) \xrightarrow{PT} (-\zeta_0, -\boldsymbol{\zeta}).$$

PT-conservation means equality of the amplitudes

$$A(p_1, \zeta_1, p_2, \zeta_2; p_3, \zeta_3, p_4, \zeta_4) = A(p_3, -\zeta_3, p_4, -\zeta_4; p_1, -\zeta_1, p_2, -\zeta_2). \tag{3.36}$$

In terms of the S-matrix elements this means

$$S_{ab} = S_{\tilde{b}\tilde{a}}, \tag{3.37}$$

where $\sim$ denotes spin flip.

Instead of the spin variables $\zeta_1, \zeta_2, \ldots, \zeta_n$ we can describe matrix elements by another set of quantum numbers, namely, by the *moduli* of the relative angular momenta, $\mathbf{J}_{ik}^2$, and the total angular momentum, $\mathbf{M}^2, M_z$. Then the matrix elements are independent of M_z, the projection of the total angular momentum, since physics cannot depend on an arbitrary choice of the z–axis. Therefore, in such a basis the PT-invariant S-matrix is symmetric:

$$S_{ab} = S_{ba}. \tag{3.38}$$

3.3 Causality and unitarity

It is usually assumed that the amplitudes of real processes must satisfy the conditions of unitarity and causality. Let us consider in detail what these conditions mean, and what restrictions on the amplitudes they lead to.

3.3.1 Causality

Consider the Compton scattering process

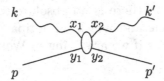

neglecting the spins for the sake of simplicity.

From the point of view of causality we are interested in the dependence on the positions of the points x_1, x_2, and we write the amplitude in the form

$$F = \int e^{ik'x_2 - ikx_1} f_{pp'}(x_1, x_2) d^4x_1 d^4x_2, \qquad (3.39)$$

where

$$f_{pp'}(x_1, x_2) = \int e^{ip'y_2 - ipy_1} f(x_1, x_2; y_1, y_2) d^4y_1 d^4y_2. \qquad (3.40)$$

Due to translational invariance (i.e. the homogeneity of space-time) the amplitude $f(x_1, x_2; y_1, y_2)$ depends only on the *differences* between the coordinates. Therefore, under the replacement

$$x_i = x_i' + a, \quad y_i = y_i' + a \qquad (3.41)$$

the function $f_{pp'}$ in (3.40) transforms as follows:

$$f_{pp'}(x_1, x_2) = e^{i(p'-p)a} f_{pp'}(x_1', x_2'), \qquad (3.42)$$

and can be written as

$$f_{pp'}(x_1, x_2) = e^{i(p'-p)\frac{x_1+x_2}{2}} \tilde{f}_{pp'}(x_2 - x_1). \qquad (3.43)$$

Then, introducing the variables $x_{21} = x_2 - x_1$ and $\frac{1}{2}(x_1 + x_2)$, we obtain for the amplitude F in (3.39)

$$F = \int e^{i(k'-k)\frac{x_1+x_2}{2} + i(k'+k)\frac{x_1-x_2}{2}} \times e^{i(p'-p)\frac{x_1+x_2}{2}} \tilde{f}_{pp'}(x_{21}) \, d^4\frac{x_1 + x_2}{2} d^4x_{21}.$$

Integration over the sum $(x_1 + x_2)/2$ gives

$$F = (2\pi)^4 \delta(p + k - p' - k') f(k, k'; p, p'),$$

where

$$f(k, k'; p, p') = \int e^{i \frac{(k+k')}{2} x_{21}} \tilde{f}_{pp'}(x_{21}) d^4 x_{21}. \qquad (3.44)$$

Conservation of the energy–momentum is as usual a result of the translational invariance. The δ-function in the amplitude arises due to space-time homogeneity.

What is causality in terms of the scattering amplitude in (3.44)? It tells us that the regions of integration

$$x_{20} < x_{10} \qquad \text{and} \qquad x_{12}^2 < 0$$

should not contribute to the amplitude. Physically this means that if a beam of particles scatters on a target, secondary particles cannot be emitted before the projectile hits the target. Hence, $\tilde{f}_{pp'}(x_{21})$ should have the form

$$\tilde{f}_{pp'}(x_{21}) = \theta(x_{20} - x_{10}) \theta(x_{21}^2) \varphi(x_{21}) + \varphi'(x_{21}), \qquad (3.45)$$

where the contribution to the amplitude (3.44) of the function $\varphi'(x_{21})$ vanishes after integration over x_{21}. Such an additional term is allowed. (The fact that (3.44) looks like the Fourier transform, does not necessarily imply $\varphi'(x_{12}) \equiv 0$. Indeed, we are considering on-mass-shell particles, $k_0 = \sqrt{\mathbf{k}^2}$, $k_0' = \sqrt{\mathbf{k}'^2}$, so that the momentum $k + k'$ has only three independent components rather than four.)

Let us assume that the function $\tilde{f}_{pp'}(x_{12})$ has the form (3.45), i.e. causality is in place. We specialize to the case of forward scattering when $k \simeq k'$, $p \simeq p'$, and choose $\mathbf{k}$ to be collinear to the z-axis. Then from (3.44) we have

$$f(k, k; p, p) = \int d^4x \, e^{ik_0 x_0 - i k_z z} \tilde{f}_p(x) = \int d^4x \, e^{ik_0(x_0 - z)} \tilde{f}_p(x) \equiv f(k_0). \qquad (3.46)$$

The representation of the function $\tilde{f}_{pp'}(x)$ in the form (3.45) means that contribution to the integral in (3.46) comes from the region

$$x_0 > 0, \quad x_0^2 - z^2 > x^2 + y^2 > 0 \qquad \Longrightarrow \qquad x_0 - z > 0.$$

How does the latter inequality affect the properties of the amplitude $f(k_0)$ as a function of energy? We wrote our amplitude for *real* (positive) values of k_0. However, once the integral is well defined (converges) on the real axis, it will converge even better in the *upper* half-plane $\text{Im } k_0 > 0$ of the complex variable k_0, due to $x_0 - z > 0$. In other words, causality leads to analyticity of the amplitude in the upper half-plane. (This conclusion may be reversed: if the amplitude is analytic, it can be expressed in the form (3.46).)

Causality is the real reason why all the amplitudes we deal with in a field theory are analytic in momenta: they are either given by explicitly analytic formulae, or expressed in terms of series that have to have definite analytic properties.

Thus, if $f(k_0)$ has a singularity in the upper half-plane, it cannot be causal. How about the behaviour at large k_0, $|k_0| \to \infty$?

In this limit, the dominant contribution to the integral (3.46) comes from the region $x_0 - z \simeq 0$. If $\tilde{f}(x)$ is singular at zero, $f(k_0)$ can increase with the growth of k_0. If, for example,

$$\tilde{f}(x) \sim \delta(x_0 - z),$$

then

$$f(k_0) \to \text{const.}$$

If $f(x)$ is even more singular,

$$\tilde{f}(x) \sim \delta'(x_0 - z),$$

then

$$f(k_0) \to k_0,$$

etc. This means that if $\tilde{f}(x)$ has the form

$$\tilde{f}(x) = \sum_n^N C_n \delta^{(n)}(x_0 - z), \tag{3.47}$$

then

$$f(k_0) = \sum_n^N C'_n k_0^n = \mathcal{O}(k_0^N). \tag{3.48}$$

Thus, we cannot ban a polynomial increase of $f(k_0)$ at infinity. Why not faster? Such a function, corresponding to an essential singularity at infinity, contains an infinite number of derivatives of the δ-function in (3.47). In this case we would not be able to guarantee that its x-space image is causal, that is, $\tilde{f}(x) = 0$ for $x_0 < z$. For example, the infinite series

$$\tilde{f}(x) = \sum_{n=0}^{\infty} a^n \delta^{(n)}(x_0 - z) = \theta(x_0 - z + a) \qquad (a > 0)$$

sum up into the obviously non-causal function,

$$\tilde{f}(x) = 1 \qquad \text{for} \quad x < z \qquad (x > z - a).$$

We consider below only polynomially bounded functions

$$f(k_0) < k_0^N \quad \text{if} \quad k_0 \to \infty. \tag{3.49}$$

This corresponds to no more than a finite number of $\delta^{(n)}$-functions in (3.47) and *guarantees* causality. We call polynomially bounded, regular in the upper half-plane functions $f(k_0)$ causal. From a formal point of view we did not prove that we *must* impose the polynomial restriction (3.49) on the growth rate to ensure causality, but such a claim seems to be relatively well founded.

Indeed, how would we verify causality experimentally? The incoming particles (photons in our case) are described by the wave function

$$\Psi(x) = \int e^{-ik_0(x_0-z)} C(k_0) \, dk_0. \tag{3.50}$$

If at $x_0 = 0$ we place the front of the wave packet at $z = a$, then the function $C(k_0)$ should be such that

$$\Psi(x) = 0 \quad \text{for} \quad z - x_0 > a, \tag{3.51}$$

i.e. the photons cannot reach the observation point z faster than with the speed of light. The condition (3.51) is valid if $C(k_0)$ has no singularities in the upper half-plane and behaves as $e^{-ik_0 a}$ on the large circle. (In this case the contour can be closed in the upper half-plane provided $x_0 - z + a < 0$, resulting in $\Psi(x) = 0$.)

The photons scattered forward are described by the wave function

$$\Psi'(x) = \int e^{-ik_0(x_0-z)} f(k_0)C(k_0) \, dk_0. \tag{3.52}$$

Causality means that after the scattering $\Psi'(x)$ also should vanish for $z - x_0 > a$. This is obviously true for a *causal* $f(k_0)$. If, however, $f(k_0)$ were growing exponentially, for example $f(k_0) \sim e^{-ik_0 c}$, $(c > 0)$ on the large circle, then $\Psi'(x)$ would vanish only for $z - x_0 > a + c$ but not for $z - x_0 > a$.

Polynomial restriction is not a matter of definition but a necessity: to verify causality we need to be able to close the contour in the upper half of the complex k_0-plane when $x_0 < z - a$.

3.3.2 Analytic properties of the Born amplitudes

Are our amplitudes analytic? The diagrams describing Compton scattering in the Born approximation are

$$F = \quad$$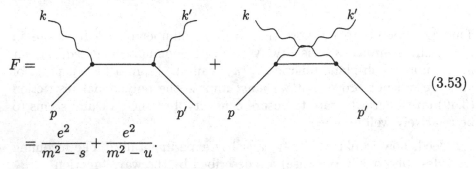

$$\qquad\qquad (3.53)$$

$$= \frac{e^2}{m^2 - s} + \frac{e^2}{m^2 - u}.$$

In the rest frame of the electron we have

$$s = (p + k)^2 = m^2 + \lambda^2 + 2mk_0,$$

$$u = (p - k')^2 = m^2 + \lambda^2 - 2mk_0' = m^2 + \lambda^2 - 2mk_0,$$

where we have introduced a small mass λ for the photons and used that for forward scattering $k_0' = k_0$. The amplitude has two poles,

$$\left.\begin{array}{l} m^2 - s = -\lambda^2 - 2mk_0 = 0 \\ m^2 - u = -\lambda^2 + 2mk_0 = 0 \end{array}\right\} \implies k_0 = \pm\frac{\lambda^2}{2m}. \qquad (3.54)$$

However, the physical amplitude is defined on the real axis where $k_0 \geq \lambda$ (the bold line in Fig. 3.1) while the poles are located at the points where $|k_0| < \lambda$, outside the physical region.

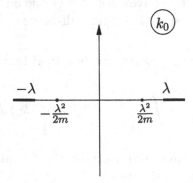

Fig. 3.1

Thus the Born amplitude possesses the correct analytical properties (has no singularities in the upper half-plane, and decreases at $|k_0| \to \infty$) and therefore respects causality.

How could this happen? Does it not contradict the fact that, as we know, the Green function $G(x_2 - x_1)$ that enters the coordinate space expression for the amplitude does not vanish for the 'wrong' time sequence, $x_{20} < x_{10}$ (and, therefore, outside the light-cone, $(x_2 - x_1)^2 < 0$)?

The second ('crossed', or 'u-channel') diagram does not pose a problem: whatever the sign of the time difference $x_{20} - x_{10}$, one of the incoming particles (either the electron p or the incident photon k) participates in the earliest interaction, so that both regions are perfectly causal.

In the first ('s-channel') diagram, however, the region $x_{20} - x_{10} < 0$ is anti-causal: in this case the final particles are created before the initial ones entered an interaction.

Let us calculate this amplitude explicitly, starting from the coordinate representation, to see what has happened.

$$F = \int d^4x_1 d^4x_2 \, e^{i(k'+p')x_2 - i(k+p)x_1} G(x_2 - x_1)$$

$$= (2\pi)^4 \delta(p + k - p' - k') \int d^4x_{21} \, e^{i(k+p)x_{21}} G(x_{21}), \tag{3.55}$$

where

$$G(x_{21}) = \int \frac{d^4q}{(2\pi)^4 i} e^{-iqx_{21}} \frac{1}{m^2 - q^2 - i\varepsilon}. \tag{3.56}$$

The integral over x_{21} in (3.55) selects a definite Fourier component with $q = k + p$ of the Green function (3.56):

$$F \propto \int d^4q \delta(k + p - q) \frac{1}{m^2 - q^2 - i\varepsilon} = \frac{1}{m^2 - (k+p)^2 - i\varepsilon}.$$

Since $q_0 = p_0 + k_0 > 0$, the left pole $q_0 = -\sqrt{m^2 + \mathbf{q}^2}$ of the function $1/(m^2 - q^2)$ effectively does not contribute to the amplitude. Since the denominator does not vanish, the sign of $i\varepsilon$ becomes unimportant. This means that we can safely move the left pole from the upper into the lower half-plane, which is equivalent to replacing the Feynman Green function (3.56) by the *retarded* one, $G_R(x_{21}) \propto \vartheta(x_{20} - x_{10})$.

The denominator of the Green function in the first diagram might vanish in the physical region of momenta if the intermediate particle were *heavier* than the electron: $e + \gamma \to \tilde{e}$ ($\tilde{m} > m + \lambda$). In this case the Feynman $i\varepsilon$ prescription would be essential and, pushing the pole at $k_0 \simeq (\tilde{m}^2 - m^2)/2m > \lambda$ into the lower half-plane, would guarantee the causal time sequence of the reaction. (In the second diagram the denominator may vanish only if the target particle is unstable, i.e. can decay spontaneously into two real particles.)

Hence, in the calculation of the s-channel Compton scattering amplitude, the Feynman Green function for the virtual intermediate electron may be replaced by the retarded Green function which vanishes for $x_{20} < x_{10}$. This is the reason why the Born amplitude has correct analytic properties. This does not mean, however, that one is allowed to introduce retarded Green functions instead of the Feynman Green functions everywhere, because this would lead to incorrect results for the scattering amplitudes, in particular in other channels.

3.3.3 Scattering amplitude as an analytic function

We have considered forward Compton scattering and shown that, due to causality, the amplitude is analytic in the upper half-plane of k_0. In the Mandelstam plane $s+t+u = 2m^2+2\lambda^2$ and $s = (m+\lambda)^2$ is the beginning of the physical region of our process (s-channel).

Recall that we have already used analyticity of the scattering amplitudes in discussion of the connection between spin and statistics in Section 2.5 in order to perform analytic continuation into the t- and u-channels. The path of the analytic continuation from the s-channel into the u-channel, marked by the arrow on the Mandelstam plane,

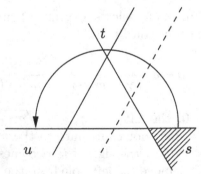

corresponds to the continuation in the k_0 plane shown in Fig. 3.2.

We see that, unlike non-relativistic quantum mechanics, in the relativistic case the region of negative k_0 also corresponds to a physical process: $s \to u$ is equivalent to $k_0 \to -k_0$. This is one of the main differences between non-relativistic and relativistic quantum theories. Analyticity (which is due to causality) makes it possible to continue the amplitude from channel to channel or in the k_0 plane.

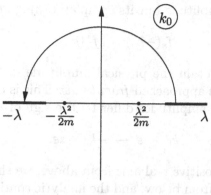

Fig. 3.2

In the simplest case we have considered, the Born amplitude had only two pole singularities and was real on the real axis. As we will show below from the *unitarity* condition, the true amplitude has to be *complex* in the physical regions of momenta shown by bold lines. The end points $k_0 = \pm\lambda$, or $s, u = (m + \lambda)^2$, are actually *branch points* so that the lines themselves are *cuts* of an analytic function. The physical s-channel amplitude is equal to the limit of this analytic function at the upper boundary of the right cut:

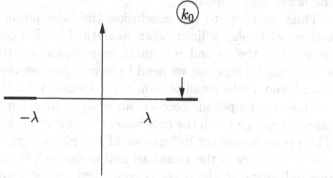

Thus, the s-channel amplitude as defined in (3.46) is analytic in the upper half-plane k_0, and hence it may be continued to negative k_0. But will we obtain an amplitude of a new physical process in this way? Apparently, to perform the analytic continuation, it is sufficient to substitute $-k_0$ for k_0. However, repeating the calculation leading to (3.46) directly in the u-channel, we obtain

$$ f_u = \int d^4x \, e^{ik_0'(x_0 - z)} f(x), $$

with $k_0' > 0$, the positive energy of the real incident photon. This tells us that by replacing $k_0 = -k_0'$ in the s-channel amplitude we would obtain not the u-channel amplitude but its complex conjugate:

$$f_s(-k_0') = f_u^*(k_0').$$

Hence, in order to obtain the physical amplitude in the u-channel, the real axis should to be approached *from below*. This is quite natural, since the arguments of the amplitudes differ by the sign of k_0:

$$k_0 + i\varepsilon \;\to\; -k_0 - i\varepsilon.$$

If we approach the positive real axis from above, we should approach the negative real axis k_0 from below, and the analytic continuation path looks as follows:

Repeating the causality considerations above but for the u-channel we would find that the physical amplitude may be analytically continued into the *lower* half-plane of k_0.

Thus we come to the conclusion that the physical amplitude in the u-channel is also a limit of an analytic function on the real axis. (To prove that the s- and u-channel amplitudes are the two limits of the same analytic function we need to have a gap between the cuts. We have introduced a fake small photon mass λ exactly for this reason.)

Thus, the amplitude becomes an analytic function in the entire complex plane of energy (with the cuts along the physical regions on the real axis). The upper and lower half-planes of the photon energy k_0 correspond to the amplitudes in the s-channel and u-channel, respectively. Going from one half-plane to the other is equivalent to the analytic continuation of the amplitude between the two channels.

3.3.4 Unitarity

The S-matrix was introduced in Section 1.11 as an operator that describes transition of the system which in the remote past was in the state

$$\Psi = \begin{pmatrix} \Psi_1 \\ \Psi_2 \\ \cdot \\ \cdot \\ \cdot \end{pmatrix},$$

into the state

$$\Psi' = S\Psi$$

in the distant future.

Matrix elements of the S-matrix S_{ba} are the transition amplitudes from the state $|a\rangle_{t=-\infty}$ to the state $|b\rangle_{t=+\infty}$. The initial and final state wave functions can be written as

$$\Psi = \sum_a |a\rangle \Psi_a ,$$

$$\Psi' = \sum_a S|a\rangle \Psi_a = \sum_{ab} |b\rangle \langle b|S|a\rangle \Psi_a \equiv \sum_b |b\rangle \Psi'_b \qquad (3.57)$$

where

$$\Psi'_b = \sum_a \langle b|S|a\rangle \Psi_a \equiv \sum_a S_{ba} \Psi_a. \qquad (3.58)$$

The last expression nicely demonstrates that the matrix elements S_{ba} are just the transition amplitudes.

Norm of any state does not change during the transition, and we have

$$\sum_a |\Psi_a|^2 = \sum_b |\Psi'_b|^2. \qquad (3.59)$$

This equation is valid for arbitrary initial states and immediately leads to unitarity of the S-matrix:

$$S S^\dagger = 1, \qquad (3.60)$$

or in the matrix form,

$$\sum_b S_{ab}S^\dagger_{bc} = \delta_{ac}. \qquad (3.61)$$

For the *diagonal* transitions $a = c$ the unitarity condition (3.61) means probability conservation:

$$\sum_b S_{ab}S^\dagger_{ba} = \sum_b |S_{ab}|^2 = 1,$$

and for the non-diagonal transitions $a \neq c$ it reflects orthogonality of the basis states

$$\sum_b S_{ab}S^\dagger_{bc} = 0.$$

It is convenient to represent the S-matrix in the form (compare with (1.124), (1.133))

$$S = 1 + iT. \qquad (3.62)$$

In terms of the T-matrix the unitarity condition (3.60) reads

$$1 + iT - iT^\dagger + TT^\dagger = 1,$$

i.e.

$$-i\left[T - T^\dagger\right] = T T^\dagger. \tag{3.63}$$

This last equation (3.63) is also often called the unitarity condition. In matrix form it can be written as

$$-i\left[T_{ba} - T_{ba}^\dagger\right] = \sum_c T_{bc} T_{ca}^\dagger,$$

or, since $T_{ba}^\dagger = T_{ab}^*$,

$$-i\left[T_{ba} - T_{ab}^*\right] = \sum_c T_{bc} T_{ac}^*. \tag{3.64}$$

The unitary condition has a more transparent form in PT-invariant theories (like quantum electrodynamics). In such a theory

$$T_{ab} = T_{\tilde{b}\tilde{a}},$$

where $|\tilde{b}\rangle$, $|\tilde{a}\rangle$ are the states with spins opposite to those of the states $|b\rangle$, $|a\rangle$. As we have discussed in Section 3.2.1, by choosing a basis in which the states are described by total angular momenta rather than spin projections, one can make the T-matrix symmetric (see (3.38)):

$$T_{ab} = T_{ba},$$

and (3.64) takes the form

$$-i\left[T_{ab} - T_{ab}^*\right] = \sum_c T_{ac} T_{cb}^*,$$

i.e.

$$\mathrm{Im}\, T_{ab} = \frac{1}{2} \sum_c T_{ac} T_{cb}^*. \tag{3.65}$$

We see that the scattering amplitude cannot be real, so that our Born amplitudes, being real, cannot be correct!

In particular, the unitarity condition for the forward scattering $a = b$ has an especially simple form,

$$\mathrm{Im}\, T_{aa} = \frac{1}{2} \sum_c |T_{ac}|^2, \tag{3.66}$$

and is called in this case the optical theorem. In terms of the forward scattering amplitude,

$$T_{aa} \propto A(\theta = 0),$$

the optical theorem simply states that the imaginary part of the forward scattering amplitude in the state a is proportional to the total cross section:

$$\sum_c |T_{ac}|^2 \propto \sigma_{tot}^{(a)}.$$

We see once again that due to the unitarity condition the scattering amplitudes in the physical region should be complex.

3.3.5 Born amplitudes and unitarity

Let us return to the Compton effect for a photon (with a small mass λ) considered above in Section 3.3.2. The physical regions in the Mandelstam plane are shaded in Fig. 3.3.

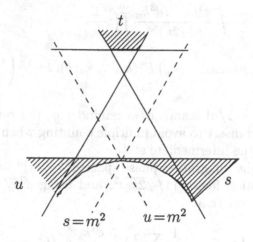

Fig. 3.3

The poles of the amplitude lie on the dashed lines corresponding to $s = m^2$, $u = m^2$.

The structure of singularities in the complex k_0 plane (two poles of the Born amplitude and two cuts along the physical regions of the s- and u-channels) is shown in Fig. 3.1.

We would like to figure out whether there is any relation between the Born amplitude and the unitarity condition.

First of all, let us write the general unitarity condition (3.65) in a more explicit form. We had (see (1.133))

$$T_{ab} = (2\pi)^4 \delta \left(\sum_a p - \sum_b p' \right) F_{ab} \prod_{i \in a} \frac{1}{\sqrt{2p_{0i}}} \prod_{i' \in b} \frac{1}{\sqrt{2p'_{0i'}}}, \qquad (3.67)$$

where p_i and $p_{i'}$ stand for the sets of the initial and final particle momenta ($i \in a$ and $i' \in b$, respectively) and F_{ab} is that very Lorentz invariant amplitude for which we have been drawing and calculating Feynman diagrams above.

Inserting (3.67) into (3.65) we obtain

$$\text{Im}\, F_{ab} = \frac{1}{2} \sum_c F_{ac} F_{cb}^* (2\pi)^4 \delta \left(\sum_{i \in a} p_i - \sum_{j \in c} k_j \right) \prod_{\ell \in c} \frac{1}{2k_{0\ell}}. \qquad (3.68)$$

The symbol $\sum_c$ implies an integration over the momenta k_ℓ of the real particles in the intermediate state c, which momenta are arbitrary (modulo the conservation law supplied by the δ-function). Writing these integrals explicitly, we have

$$\text{Im}\, F_{ab} = \frac{1}{2} \sum_c \int \frac{d^3 k_1 \dots d^3 k_n}{n!\,(2\pi)^{3n}} \prod_{i=1}^n \frac{1}{2k_{0i}}$$

$$\times F(a; k_1, \dots k_n)\, F^*(k_1, \dots k_n; b)(2\pi)^4 \delta \left(\sum_a p - \sum_{i=1}^n k_i \right),$$
$$\qquad (3.69)$$

where the factor $1/n!$ stands as a reminder of the combinatorial factor that one should insert to avoid multiple counting when identical particles are present in the intermediate state.

Combining the integration phase space $d^3 k/(2\pi)^3$ with the wave function normalization factor $(1/\sqrt{2k_0})^2$, and using $d^3 k/2k_0 = d^4 k \delta_+(k^2 - m^2)$, we finally arrive at

$$\text{Im}\, F_{ab} = \frac{1}{2n!} \sum_n \prod_{i=1}^n \left(\frac{d^4 k_i}{(2\pi)^3} \delta_+(k_i^2 - m_i^2) \right)$$

$$\times F_{ac}\, F_{cb}^* (2\pi)^4 \delta(\sum_a p - \sum_{i=1}^n k_i). \qquad (3.70)$$

We sometimes write (3.70) in the symbolic form

$$\text{Im}\, F = \sum_n F_n F_n^*.$$

In the case of forward scattering ($a \equiv b$) we observe that the right-hand side of (3.70) differs from the total cross section only by the flux factor

$$\frac{1}{4E_1 E_2 j} = \frac{1}{4p_c \sqrt{s}}$$

(with p_c the centre-of-mass momentum of colliding particles, see (1.139)) and by the factor $1/2$ before the sum. This leads to the optical theorem in the form

$$\operatorname{Im} F(s,0) = 2p_c\sqrt{s}\,\sigma_{tot}. \tag{3.71}$$

Let us return to the Born amplitude

$$= \frac{e^2}{m^2 - s - i\varepsilon}.$$

We can easily calculate its imaginary part with the help of the well known relation (1.164),

$$\operatorname{Im} \frac{e^2}{m^2 - s - i\varepsilon} = e^2\,\pi\delta(s - m^2). \tag{3.72}$$

On the other hand, the imaginary part of any amplitude is given by (3.70) and can be represented graphically as

this can be anything

The intermediate states c consist of real particles and therefore it is impossible kinematically to have a single-electron intermediate state. Nevertheless, we can formally consider the contribution of such a state to the imaginary part.

$$F_{21} =$$

This contribution is given by the expression

$$\operatorname{Im} F_{2\to 2} = \frac{2\pi}{2} \int \frac{d^4q}{(2\pi)^3}\delta_+(q^2 - m^2)\,|F_{2\to 1}|^2\,(2\pi)^4\delta(p + k - q) \tag{3.73}$$

$$= \pi\,|F_{2\to 1}|^2\,\delta_+((p + k)^2 - m^2)$$

and is equal to the imaginary part of the Born amplitude in (3.72). We see that our Born amplitude actually respects the unitarity condition in a certain sense, as its imaginary part is determined by the (kinematically forbidden) one-electron intermediate state.

We could verify that the imaginary part of the amplitude has indeed the form of the unitarity condition by considering more complicated processes in which we can get a one-electron intermediate state contribution to the imaginary part without contradicting energy–momentum conservation.

This is possible, for example, in the transition of three particles into three:

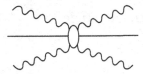

The imaginary part of this amplitude contains (among others) the graph

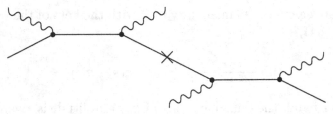

Here the electron marked by the cross × may be real (in this case it is not banned by the conservation laws).

3.3.6 How to restore perturbation theory on the basis of unitarity and analyticity, or perturbation theory without Feynman graphs

We have just proved (see (3.73)) that in a certain sense the Born amplitudes satisfy the unitarity condition. Indeed, the imaginary part of the s-channel Born Compton amplitude $e^-\gamma \to e^- \to e^-\gamma$ formally has the structure prescribed by the unitarity condition (3.70) even despite the fact that the real one-electron contribution on the right-hand side in (3.70) is kinematically banned.

Will the Born amplitude still satisfy even such a loosely interpreted unitarity condition when we increase energy?

The unitarity relation (3.70) is a non-linear equation, and which contributions are allowed on the right-hand side depends on energy. With the increase in energy the first non-vanishing contribution to the imaginary part of the Compton scattering amplitude $\mathrm{Im}\, F$ in (3.70) arises from two real particles $(e^- + \gamma)$ in the intermediate state and this contribution is given by the *product* of two Born Compton amplitudes, FF^*. Since the Born Compton scattering amplitude is of order e^2, this intermediate state generates a contribution of order e^4 to the imaginary part of the scattering amplitude, $\mathrm{Im}\, F = \mathcal{O}(e^4)$. Thus we may say that the Born amplitude is only a lowest order (e^2) approximation to the true scattering amplitude, and the higher order contributions are generated from the Born amplitude through the unitarity relation (3.70).

This observation suggests the idea of a new construction for perturbation theory, using the unitarity relation as a tool for calculating the full amplitude as a series in the coupling constant e^2.

Indeed, creation of a particle in electrodynamics can go only through the vertex that is proportional to the electric charge e. Therefore, the addition of one particle to the intermediate state (either by adding a photon or replacing one photon by two particles e^+e^-) adds a factor e^2 to the right-hand side of the unitarity equation. We see that the contribution to the imaginary part of an n-particle intermediate state is of order e^{2n}, and expansion over the number of particles in intermediate states is at the same time an expansion in powers of the charge. This connection between the number of particles in the intermediate state and power of the coupling constant leads to a scheme of the perturbation theory independent of, but equivalent to, the Feynman diagram technique.

For example, the unitarity condition for two-photon annihilation into an electron–positron pair has the form

$$\text{Im}\, F(2\gamma \to e^+e^-) = \quad\text{[diagram]}\quad + \text{Im} \quad\text{[diagram]}\quad + \cdots$$

If we want to calculate the imaginary part of this amplitude with accuracy up to e^4, we neglect all terms of order $e^n, n \geq 6$. Then it suffices to consider two-particle intermediate states, and these contributions may easily be calculated in terms of the known Born amplitudes. Iteration by iteration, we could in principle calculate imaginary parts of the amplitude to higher orders in all regions of the Mandelstam plane. There is, however, one obstacle: the imaginary part is expressed in (3.68) in terms of the *total* amplitudes of lower orders, so to have a regular perturbation theory we need a tool to calculate the real parts of the amplitudes as well. And of course, the total amplitude (equal to the sum of the real and imaginary parts) is needed first of all for a description of physical processes.

A crucial rôle in restoring the total amplitude from its imaginary part is played by analyticity. According to the Cauchy theorem, the analytic scattering amplitude for Compton scattering may be represented as

$$f(k_0) = \int \frac{dk_0'}{2\pi i}\, \frac{f(k_0')}{k_0' - k_0}, \qquad (3.74)$$

where the contour of integration is shown in Fig. 3.4:

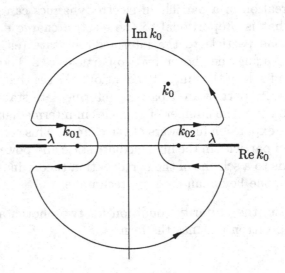

Fig. 3.4

The amplitude goes to zero at infinity, hence the integrals over the large circles vanish and the contour can be represented in the form shown in Fig. 3.5:

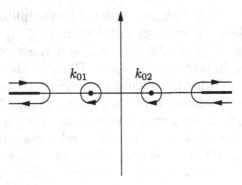

Fig. 3.5

Then

$$f(k_0) = \int \frac{dk_0'}{2\pi i} \frac{f(k_0')}{k_0' - k_0} = \frac{r_1}{k_{10} - k_0} + \frac{r_2}{k_{20} - k_0}$$
$$+ \frac{1}{\pi} \int_\lambda^\infty \frac{dk_0'}{k_0' - k_0} \operatorname{Im} f(k_0') + \frac{1}{\pi} \int_{-\infty}^{-\lambda} \frac{dk_0'}{k_0' - k_0} \operatorname{Im} f(k_0'),$$

(3.75)

since

$$\frac{1}{2i} \left\{ f(k_0' + i\varepsilon) - f(k_0' - i\varepsilon) \right\} = \operatorname{Im} f(k_0').$$

The terms in (3.75) containing r_1 and r_2 correspond to the contributions coming from the poles k_{01} and k_{02} of the Born approximation. (We have accepted here the hypothesis that no other pole singularities emerge in higher orders.)

Now we have a regular perturbation theory for the scattering amplitudes. If the imaginary part of the amplitude of some order in the coupling constant is determined from the unitarity condition, then due to analyticity the corresponding real part (the full amplitude) may be found with the help of the Cauchy theorem. Given this amplitude (together with other related amplitudes at the same order in e^2) the imaginary part of the next order may be obtained, again via unitarity, and so on.

For example, using the Born approximation for two-photon fusion into an electron–positron pair, $2\gamma \to e^+e^-$, we can restore the photon–photon scattering amplitude. Indeed,

$$\text{Im}\, F(\gamma\gamma \to \gamma\gamma) = \quad\quad\quad\quad\quad\quad\quad \tag{3.76}$$

The building blocks

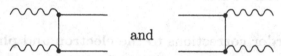

and

are just the Born amplitudes, and by substituting (3.76) into (3.75), we find the amplitude of the light-by-light scattering (the only assumption being the absence of point-like four-gamma interaction).

The problem we could face carrying out such a program is that of *convergence* of the dispersion (Cauchy) integrals in (3.75). This problem does arise, but only when the unitarity graphs have the structure of loop insertions into single-particle (electron, photon) lines or that of 'triangular' correction to the basic $ee\gamma$ vertex, and can be overcome by employing the physical particle masses and the physical coupling. (We shall return to these issues later in Chapter 5.)

This form of perturbation theory deals directly with scattering amplitudes of physical particles and avoids even mentioning their Green functions.

Having satisfied the unitarity condition in each term of a given order in e^2, we are not guaranteed, however, against trouble after attempting to *collect* the perturbative series. Indeed, as we shall see below, quantum electrodynamics *per se* formally fails at academically large energies

$$E \sim m \exp\left\{\frac{4\pi^2}{e^2}\right\} \sim m\, 10^{200}.$$

4

Radiative corrections. Renormalization

We have shown in the previous chapter how one can obtain higher order corrections to the scattering amplitudes from the Born terms using dispersion relations. There is, however, a simpler way to construct higher order multiloop amplitudes directly, namely, the method of Feynman diagrams, which we will consider below.

4.1 Higher order corrections to the electron and photon Green functions

4.1.1 Multiloop contributions to the electron Green function

Let us consider first a free charged particle.* What could happen to this free particle?

(1) The particle could propagate freely from x_1 to x_2:

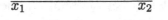

(2) The particle could emit a photon at point x_1'

Emission of a free photon is, however, banned by conservation laws. The photon can exist only for a finite time allowed by the uncertainty principle, and then it has to be absorbed by the same particle:

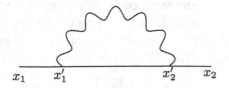

---

* We assume for the time being that only one species of charged particles exists.

174

(3) More complicated processes, like

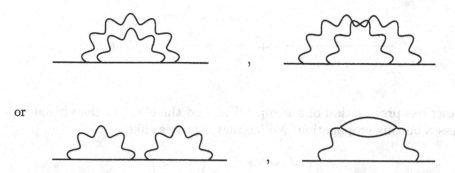

could take place.

The exact Green function describing propagation of a free particle is equal to the sum of Green functions of all such processes. Notice that each electron–positron pair is accompanied by an extra factor -1 since $\bar{v}(p) = -\bar{u}(-p)$ (see (2.49), more in Section 4.1.2).

The essence of the Feynman method for any process is to draw all topologically different graphs of all orders in the coupling constant for the process under consideration, and then sum them to obtain the respective Green functions.

Let us emphasize that one should not overcount topologically identical diagrams. For example, the graph with a closed electron–positron loop

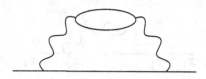

can be represented either as

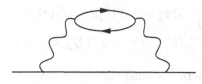

or as

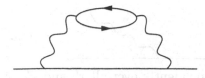

This is essentially one and the same process, so it should be taken into account only once.

The sum of the diagrams

describes propagation of a free particle, and the effect of the virtual processes on this propagation. Notice that a process like

is merely a replication of the process

Diagrams of this type can be easily taken care of. We will see later that it is convenient to consider separately the processes that do not reduce to simple replication.

First, let us formulate the correspondence rules between the higher order multiloop diagrams and the Green functions. The simplest one-loop graph

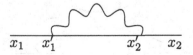

is described by the following analytic expression

$$G_2(x_2 - x_1) = e^2 \int G(x_2 - x_2')\, i\Gamma_\mu(x_2')\, G(x_2' - x_1')$$
$$\times\, i\Gamma_\nu(x_1')\, G(x_1' - x_1)\, D_{\mu\nu}(x_2' - x_1')\, d^4x_2'\, d^4x_1'\,. \tag{4.1}$$

The factor e^2 arises here because there are two vertices. As usual, each vertex in x-space (as well as in the momentum space) also contains the factor $i\gamma_\mu$

 $\to i\gamma_\mu.$

Integration goes independently over all x_1' and x_2', since virtual emission and absorption of the photon could happen independently at any moment of time and at any point in space.

In the next order of perturbation theory in the coupling constant we can consider, for example, the two-loop graph

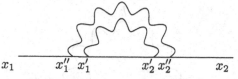

$$x_1 \qquad x_1'' \quad x_1' \qquad\qquad x_2' \quad x_2'' \qquad x_2$$

Using the same logic as above, we write for this graph

$$
\begin{aligned}
G_4(x_2 - x_1) = e^4 \int & G(x_2 - x_2'')\, i\gamma_\mu G(x_2'' - x_2')\, i\gamma_\nu G(x_2' - x_1') \\
& \times i\gamma_{\nu'} G(x_1' - x_1'')\, i\gamma_{\mu'} G(x_1'' - x_1) D_{\mu\mu'}(x_2'' - x_1'') \quad (4.2) \\
& \times D_{\nu\nu'}(x_2' - x_1')\, d^4 x_1''\, d^4 x_1'\, d^4 x_2'\, d^4 x_2''.
\end{aligned}
$$

Now we can formulate a general rule, how to write an analytic expression for an arbitrary multiloop diagram for the electron Green function. First, we put down a product of all free Green functions corresponding to all lines in the diagram starting with the last one. Then we integrate over positions of all interaction points. We also have to make contractions of all repeating indices $\mu, \mu', \ldots$ of the photon Green functions and vertices.

Let us derive Feynman rules for multiloop diagrams in the momentum representation. To this end we substitute Fourier representations for the free electron and photon Green functions (see (1.85) and (2.56)),

$$
\begin{aligned}
G(x) &= \int \frac{d^4 p}{(2\pi)^4 i} \frac{e^{-ipx}}{m - \hat{p}}, \\
D_{\mu\nu}(x) &= g_{\mu\nu} \int \frac{d^4 k}{(2\pi)^4 i} \frac{e^{-ikx}}{k^2},
\end{aligned} \qquad (4.3)
$$

in the coordinate space expressions (4.1) or (4.2) for the higher order diagrams. All coordinate dependence in the Fourier representations in (4.3) is in the exponents, and integrations over coordinates in (4.1) and (4.2) become trivial after this substitution.

For example, for (4.1) we immediately obtain

$$
\begin{aligned}
G_2(x_2 - x_1) = e^2 \int & \frac{d^4 p_1 d^4 p_2 d^4 p_3 d^4 k}{[(2\pi)^4 i]^4} \frac{1}{m - \hat{p}_2} i\gamma_\mu \frac{1}{m - \hat{p}_3} i\gamma_\nu \frac{1}{m - \hat{p}_1} \frac{g_{\mu\nu}}{k^2} \\
& \times \int e^{-ip_2 x_2} e^{-i(-p_2 + p_3 + k)x_2'} e^{-i(-p_3 + p_1 - k)x_1'} e^{ip_1 x_1} d^4 x_1' d^4 x_2'.
\end{aligned}
$$

$$(4.4)$$

The integral

$$
\int e^{-ix_2'(-p_2 + p_3 + k) - ix_1'(-p_3 + p_1 - k)} d^4 x_1' d^4 x_2'
$$

$$
= (2\pi)^4 \delta(p_3 + k - p_2)(2\pi)^4 \delta(p_1 - k - p_3),
$$

simply demonstrates that in momentum representation four-momentum is conserved in each vertex (see Fig. 4.1).

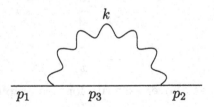

Fig. 4.1

Momentum δ-functions lift integrals over p_2 and p_3 in (4.4), and we obtain

$$G_2(x_2 - x_1) = \int \frac{d^4 p_1}{(2\pi)^4 i} e^{-ip_1(x_2 - x_1)}$$

$$\times \int \frac{d^4 k}{(2\pi)^4 i} e^2 \frac{1}{m - \hat{p}_1} \gamma_\mu \frac{1}{m - (\hat{p}_1 - \hat{k})} \gamma_\mu \frac{1}{m - \hat{p}_1} \frac{1}{k^2}. \tag{4.5}$$

Comparing this expression with the general Fourier representation for the diagram in Fig. 4.1 ($p = p_1 = p_2$)

$$G_2(x_2 - x_1) = \int \frac{d^4 p}{(2\pi)^4 i} e^{-ip(x_2 - x_1)} \tilde{G}_2(p),$$

we see that the one-loop Green function $\tilde{G}_2(p)$ in momentum space has the form

$$\tilde{G}_2(p) = \frac{e^2}{m - \hat{p}} \left(\int \frac{d^4 k}{(2\pi)^4 i} \gamma_\mu \frac{1}{m - \hat{p} + \hat{k}} \gamma_\mu \frac{1}{k^2} \right) \frac{1}{m - \hat{p}}. \tag{4.6}$$

Integration over momentum k of the intermediate photon survived in the one-loop correction in (4.6)

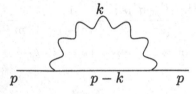

to the free Green function. This happened because four-momentum conservation in each vertex of the diagram still does not fix the momentum of the intermediate state, and the virtual photon can have an arbitrary momentum.

Green functions corresponding to any diagram may be calculated in the same way. For instance, diagrams for corrections of order e^4 to the electron Green function look like

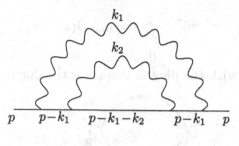

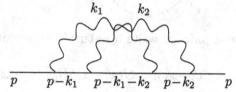

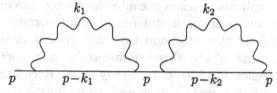

Analytically the Green function for the second graph has the form

$$
\tilde{G}_4(p) = e^4 \frac{1}{m - \hat{p}} \left(\int \frac{d^4k_1 d^4k_2}{[(2\pi)^4 i]^2} \gamma_\mu \frac{1}{m - \hat{p} + \hat{k}_2} \gamma_\nu \frac{1}{m - \hat{p} + \hat{k}_2 + \hat{k}_1} \right.
$$
$$
\left. \times \gamma_\mu \frac{1}{m - \hat{p} + \hat{k}_1} \gamma_\nu \frac{1}{k_1^2} \frac{1}{k_2^2} \right) \frac{1}{m - \hat{p}},
$$

(4.7)

and similar expressions can be written for the other graphs.

4.1.2 Multiloop contributions to the photon Green function

Let us consider propagation of a free photon. The only thing that could happen to a free photon is that it would decay into an electron–positron pair for a short time. Nothing else is possible since the photon can interact only through the triangle electron–photon vertex. Thus, only processes like can contribute to the photon Green function.

What are the analytic expressions for these diagrams? As we know, the lowest order photon Green function $g_{\mu\nu}D(x_2 - x_1)$ corresponds to the diagram

$$= g_{\mu\nu}\, D(x_2 - x_1). \qquad (4.8)$$

Using our experience with the photon loops, for the diagram

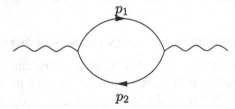

with an electron–positron pair we write a natural expression

$$e^2 \int D_{\mu\mu'}(x_2-x_2')i\gamma_{\mu'}G(x_2'-x_1')i\gamma_{\nu'}G(x_1'-x_2')D_{\nu'\nu}(x_1'-x)d^4x_1'd^4x_2'. \qquad (4.9)$$

An additional complication arises because the photon can create an intermediate electron and positron in different, though correlated spin states. Hence, we should additionally sum over all allowed intermediate spin states in (4.9). First of all, this summation means calculation of the trace of the expression in (4.9), but there is more. Recall that a (virtual) photon transition into a real electron and a real positron is described by the vertex

$$= e\bar{u}(p_1)\gamma_\mu v(p_2).$$

So, if in the internal part of the diagram

the particles were real, then the amplitude would have the form

$$A \propto \sum_{\lambda_1\lambda_2} \bar{u}^{\lambda_1}(p_1)\gamma_\mu v^{\lambda_2}(p_2)\bar{v}^{\lambda_2}(p_2)\gamma_\nu u^{\lambda_1}(p_1)$$

$$= -\sum_{\lambda_1\lambda_2} \bar{u}^{\lambda_1}(p_1)\gamma_\mu u^{\lambda_2}(-p_2)\bar{u}^{\lambda_2}(-p_2)\gamma_\nu u^{\lambda_1}(p_1). \qquad (4.10)$$

The minus sign arose here because $\bar{v}(p) = -\bar{u}(-p)$ (see (2.49)). Using the completeness relation (2.49)

$$\sum_\lambda u_\alpha^\lambda(p)\bar{u}_\beta^\lambda(p) = (\hat{p}+m)_{\alpha\beta},$$

we reduce summation in (4.10) to calculation of the same trace as in (4.9) but with an opposite sign. Thus if we insist on the correspondence between (4.9) for real intermediate particles and (4.10) for virtual particles in the intermediate state, we have not only to take the trace in (4.9) but additionally to multiply it by minus one.

It is easy to realise that this is a general rule, and an extra factor (-1) accompanies each closed electron–positron loop in an arbitrary graph.

In the same way as for the electron Green functions above, we can derive the momentum space representation for the photon Green functions. For example, the one-loop contribution to the photon Green function in momentum space looks like

$$
D_{\mu\nu}^{(2)}(k) = {}^k\!\!\!\!\!\underset{p-k}{\overset{p}{\rule{0pt}{0pt}}}\!\!\!\!\!\!\!{}^k
$$

$$(4.11)$$

$$= -\frac{e^2}{k^2}\left(\mathrm{Tr}\int\frac{d^4p}{(2\pi)^4 i}\gamma_\mu\frac{1}{m-\hat{p}}\gamma_\nu\frac{1}{m-\hat{p}+\hat{k}}\right)\frac{1}{k^2}.$$

The only complication in comparison with the Feynman rules for the loops made of charged scalar particles is that now we have to take the trace over the spinor indices, and write an extra minus sign for every virtual electron–positron pair.

A real photon with $k^2 = 0$ cannot decay into two real particles without violating energy–momentum conservation. In the virtual process

the photon is off mass shell $k^2 \neq 0$, and the electron also is off the mass shell $p_0 \neq \sqrt{\mathbf{p}^2 + m^2}$. In terms of non-relativistic quantum mechanics we could say that in this process a virtual photon for a short time (determined by the uncertainty relation) decays into two particles, violating energy conservation.

Higher order corrections in the Feynman diagram approach are constructed via relativistically invariant virtual processes. The virtual particles in the Feynman diagrams are off mass shell. This should be compared with the ordinary quantum mechanical perturbation theory where there is no energy conservation for the intermediate states. The Feynman method is in principle equivalent but much more convenient than the old non-covariant perturbation theory because it is explicitly relativistically invariant.

4.2 Renormalization of the electron mass and wave function

Let us consider exact Green functions for free charged particles in more detail.[†] The total Green function of a free charged particle with mass m_0 is given by the sum of terms which correspond to all possible processes of photon emission and absorption

$$ G \equiv \quad\rule{2cm}{0.4pt}\quad = \quad\rule{1.5cm}{0.4pt}\quad + \quad + \quad + \cdots $$

Some of the diagrams on the right-hand side, which contain subdiagrams connected by one electron line, describe simply repetitions of the same fluctuations

These fluctuations are only weakly correlated, and might be separated by large time intervals. On the other hand, fluctuations corresponding to one-particle irreducible diagrams, like

are strongly correlated and occur via a very short time interval. In order to separate weakly and strongly correlated fluctuations, we introduce the notion of self-energy of a particle. The self-energy is a sum of all one-particle irreducible diagrams (i.e. a sum of diagrams which do not contain weakly correlated fluctuations)

$$ -\Sigma(p) = \quad + \quad + \quad + \quad + \cdots = \quad $$

Self-energy $\Sigma(p)$ contains only fluctuations which take place in a short time interval. All other fluctuations which contribute to the total electron Green function may be obtained simply by replicating the self-energy graphs. Hence, the total Green function can be written as[‡]

$$ G(p) = \frac{}{G_0} + \frac{}{G_0 \quad -\Sigma} + \frac{}{G_0} + \cdots $$

[†] Below we will call the mass which is written in the free electron propagator m_0 instead of m.

[‡] From now on bare Green functions of the type (4.3) will carry a subscript 0.

Analytically the series for the electron Green function has the form

$$G(p) = \frac{1}{m_0 - \hat{p}} + \frac{1}{m_0 - \hat{p}} [-\Sigma(p)] \frac{1}{m_0 - \hat{p}}$$
$$+ \frac{1}{m_0 - \hat{p}} [-\Sigma(p)] \frac{1}{m_0 - \hat{p}} [-\Sigma(p)] \frac{1}{m_0 - \hat{p}} + \cdots$$
$$= \frac{1}{m_0 - \hat{p}} \left[1 + [-\Sigma(p)] \frac{1}{m_0 - \hat{p}} \right. \tag{4.12}$$
$$\left. + \left([-\Sigma(p)] \frac{1}{m_0 - \hat{p}} \right)^2 + \cdots \right].$$

This is a geometric series, and its sum may be easily calculated in terms of the self-energy Σ

$$G(p) = \frac{1}{m_0 - \hat{p}} \frac{1}{1 + \Sigma(p) \frac{1}{m_0 - \hat{p}}} = \frac{1}{m_0 - \hat{p} + \Sigma(p)}. \tag{4.13}$$

In zeroth order approximation the electron self-energy vanishes, and our new formula reproduces the well known expression for the free electron Green function

$$G_0(p) = \frac{1}{m_0 - \hat{p}} = \frac{m_0 + \hat{p}}{m_0^2 - p^2}.$$

The parameter m_0 in this expression should be interpreted as the mass of the particle, since $G_0(p)$ has a pole at the point $p^2 = m_0^2$. Indeed, particle propagation in coordinate space from x_1 to x_2 is described by the Fourier transform of the momentum space Green function (see (2.56)), and for the free Green function we have

$$G_0(x_{21}) = \int \frac{d^4p}{(2\pi)^4 i} e^{ipx_{21}} \frac{m_0 + \hat{p}}{m_0^2 - p^2},$$

where $x_{21} = x_2 - x_1$.

Calculating this integral via residues, we obtain

$$G_0 \sim \int d^3p \, e^{-ip_0 t_{21} + i\mathbf{p} \cdot \mathbf{r}_{21}},$$

where $p_0 = \sqrt{m_0^2 + \mathbf{p}^2}$. This relativistic dependence of energy on momentum just means that the Green function describes propagation of a particle with mass m_0. Moreover, at large times $t_2 \to \infty$, the only contribution to the integral comes from the pole term. If there were no pole, the integral would be zero due to fast oscillations of the exponent, and we would not observe any particles at all. It is the existence of the pole which makes the integral non-vanishing, and provides the proper relativistic relation between energy and momentum of the particle.

There is no reason to believe that after the calculation of radiative corrections the position of the pole of the exact electron Green function (4.13) would coincide with the position $p^2 = m_0^2$ of the pole of the zero order Green function G_0. Hence, the parameter m_0 has no direct relation to the physically observable electron mass, and the latter should be determined from the equation

$$G(p)|_{p^2=m^2} = \infty \qquad (4.14)$$

for the total electron Green function. We see that the true mass of a particle depends on its self-energy. The very existence of a free particle with an observable mass m puts certain restrictions on the self-energy $\Sigma(p)$.

The self-energy $\Sigma(\hat{p})$ depends on γ-matrices only through $\hat{p}$. Thus it commutes with $\hat{p}$, and the inverse Green function can be written in the form

$$G^{-1} = m_0 - \hat{p} + \Sigma(\hat{p}).$$

Equation (4.14) for the position of the physical mass is equivalent to

$$(m_0 - \hat{p} + \Sigma(\hat{p}))\, u_m(p) = (m_0 - m + \Sigma(m))\, u_m(p) = 0,$$

where $u_m(p)$ is a solution of the free Dirac equation with mass m. Then the observable mass of a particle is a real root of the equation

$$m_0 - m + \Sigma(m) = 0. \qquad (4.15)$$

The parameter m_0 is not observable, so it would be better to get rid of it, replacing it by a certain function of m. This can be easily achieved with the help of the relationship

$$G^{-1}(p) = m_0 - \hat{p} + \Sigma(p),$$

where we substitute m_0 from (4.15). Then we arrive at an expression for the exact Green function of the electron written only in terms of the observable mass

$$G^{-1}(p) = m - \hat{p} + \Sigma(\hat{p}) - \Sigma(m). \qquad (4.16)$$

Let us now turn to higher order corrections to the charged particle wave function. It is not difficult to realize that such corrections are intimately connected with the higher order corrections to the Green function.

Recall how we derived the scattering amplitudes in Sections 1.11 and 2.3. The idea was to consider Green functions corresponding to external legs of a diagram

$$A \sim \int \frac{d^4p}{(2\pi)^4 i} \frac{e^{-ip(x_1-x_1')}}{m_0^2 - p^2}(m_0 + \hat{p})$$

Then, using time ordering of the space-time points x_1, x_1', we closed the integration contour in the complex p_0-plane around the pole at the point m_0 and calculated the residue. With the help of the completeness relation (2.49)

$$\sum_\lambda u_\alpha^\lambda(p)\bar{u}_\beta^\lambda = (\hat{p} + m_0)_{\alpha\beta},$$

the Green function corresponding to an external leg of a diagram may be reduced to the form

$$-\sum_\lambda \int \frac{d^3p}{2p_0(2\pi)^3} \, e^{-ip(x_1-x_1')} \, \bar{u}^\lambda(p) \, u^\lambda(p),$$

where now $p_0 = \sqrt{\mathbf{p}^2 + m_0^2}$. Further, we simply omitted the wave functions of the free particles

$$\frac{1}{\sqrt{2p_0}}u^\lambda(p)e^{-ipx_1}, \qquad \frac{1}{\sqrt{2p_0}}e^{ipx_1'}\bar{u}^\lambda(p)$$

in the expressions for the multiparticle Green functions and the remaining factor turned out to be just the scattering amplitude (compare (1.130)–(1.131)).

The exact Green function has a pole at the position of the physical mass m instead of m_0. Let us see how the residue at the new pole changes in comparison with the residue of the free Green function at m_0. As we just explained, the residue is connected to the scattering amplitude, so we hope to find how such amplitude changes due to change of the Green function. To answer this question we expand $\Sigma(p)$ in a series near m:

$$\Sigma(p) = \Sigma(m) + (\hat{p} - m)\Sigma'(m) + (\hat{p} - m)^2\tilde{\Sigma}(p). \tag{4.17}$$

Let us write the last term in (4.17) which contains all the higher power terms in $(\hat{p} - m)$ in the form

$$[1 - \Sigma'(m)]\Sigma_c(p) \equiv (\hat{p} - m)^2\tilde{\Sigma}(p), \tag{4.18}$$

or

$$\Sigma_c(\hat{p}) = \frac{\Sigma(\hat{p}) - \Sigma(m) - (\hat{p} - m)\Sigma'(m)}{1 - \Sigma'(m)}. \tag{4.19}$$

In these terms, representation (4.16) for the exact electron Green function becomes

$$G^{-1}(p) = [m - \hat{p} + \Sigma_c(p)] \, [1 - \Sigma'(m)] \equiv [1 - \Sigma'(m)] \, G_c^{-1}(p), \tag{4.20}$$

where

$$G_c^{-1}(p) = m - \hat{p} + \Sigma_c(\hat{p}) \tag{4.21}$$

is called the renormalized electron Green function.

Near the pole $\hat{p} = m$ the function $\Sigma_c(p)$ turns to zero as at least the second power of $(\hat{p} - m)$ (see (4.18)) and, hence, we can forget about it calculating the scattering amplitude in the limit $x_1 \to \infty$. Then we obtain

$$A \propto \int \frac{d^4p}{(2\pi)^4 i} \frac{e^{-ip(x_1-x_1')}}{(m - \hat{p})[1 - \Sigma'(m)]}$$

$$\overset{t_1 \to \infty}{=\!=} \sum_\lambda \int \frac{d^3p}{2p_0(2\pi)^3} u^\lambda(p)e^{-ipx_1} \bar{u}^\lambda(p)e^{ipx_1'} \frac{1}{1 - \Sigma'(m)}$$

for the pole contribution to the Green function of the external leg. Our spinors were normalized by the condition $\bar{u}u = 2m$. If we insist now that the residue of the Green function at the particle pole is still equal to the product of the wave functions, we have to introduce new spinors

$$u' = \sqrt{Z_2}\, u, \quad \bar{u}' = \sqrt{Z_2}\, \bar{u}, \tag{4.22}$$

where

$$Z_2 = \frac{1}{1 - \Sigma'(m)}. \tag{4.23}$$

The new spinors in (4.22) are normalized by the condition $\bar{u}'u' = 2mZ_2$. Thus, the wave functions of the electron are now renormalized, and the constant Z_2 for obvious reasons is called the electron wave function renormalization constant. Physically, renormalization of the electron wave function means that the system we are considering contains photons and e^+e^- pairs in addition to the electron:

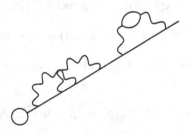

The wave function of this system has the form

$$\Psi_{\text{physical}} = \Psi_e + \Psi_{e\gamma} + \Psi_{e2\gamma} + \Psi_{e3\gamma} + \cdots + \Psi_{e^+e^-e^-} + \cdots.$$

Hence, if the wave function of the system as a whole is normalized to unity, the norm of the one-electron component Ψ_e cannot be unity any more. This norm is just the fraction of the one-electron state (bare electron) in the whole multiparticle state (physical electron).

Of course, normalization of the wave functions should not affect any observables. In the case under consideration such an observable is the cross section. The cross section contains as factors the inverse flux of the initial particles and the phase volume of the final particles, which also depend on the normalization of the wave functions. Wave functions of *physical* electrons bring in Z_2^{-1} to the flux factor (the latter being inversely proportional to the density of incoming particles). Correspondingly, the phase volume of each final *physical* electron should contain the factor Z_2. Since the cross section contains scattering amplitude *squared*, one of the square roots $\sqrt{Z_2}$ for each external leg of the amplitude will cancel with the respective factor in the flux or will be absorbed into the phase volume.

Bearing this in mind, we can simply calculate the flux and the phase volume as usual ignoring any additional factors, while including in the calculation of the scattering amplitude one square root $\sqrt{Z_2}$ for each external leg.

Thus, taking account of interactions, an additional multiplicative factor $\sqrt{Z_2}$ arises for each incoming and outgoing free electron line in the scattering amplitude:

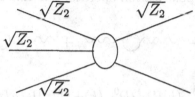

Let us summarize our discussion of the higher order corrections to the electron Green function. After summation of the higher order corrections, the electron Green function no longer coincides with the free electron Green function: both the electron mass and the electron wave function change. The observable physical electron mass m is often called renormalized mass, as opposed to the bare (or unrenormalized) mass parameter m_0. The Green function of a physical electron may be written exclusively in terms of the renormalized mass m (in place of the unobservable bare mass m_0). The residue of the Green function at the physical pole $p^2 = m^2$ can be set to unity, corresponding to propagation of *one* particle, which redefinition of the physical state brings in $\sqrt{Z_2}$ factors to the calculation of the renormalized scattering amplitudes.

4.3 Renormalization of the photon Green function

Now consider radiative corrections to the photon Green function. The exact photon Green function $D_{\mu\nu}$ has the form and on the basis of our experience with the electron Green function we would expect that the position of the pole in the exact photon Green function would shift to some non-zero value. As we know, the position of the pole is just the photon mass, and if higher order diagrams generated a finite photon mass, the photon would no longer be a photon, and our theory would not be electrodynamics. Clearly, this problem deserves further investigation. We are going to show that, due to current conservation, the position of the pole in the exact photon Green function does not shift after accounting for the radiative corrections.

Let us first introduce the photon polarization operator, defined as a sum of all diagrams which cannot be disconnected by cutting only one photon line:

$$\Pi_{\mu\nu}(k) = \quad \longleftrightarrow \quad + \quad \longleftrightarrow \quad + \cdots \equiv \quad \longleftrightarrow .$$

The photon polarization operator is similar to the electron self-energy $\Sigma(p)$. In terms of the polarization operator the exact photon Green function reads

$$\sim\!\!\sim\!\!\sim = \sim\!\!\sim\!\!\sim + \sim\!\!\sim\!\bigcirc\!\sim\!\!\sim + \sim\!\!\sim\!\bigcirc\!\sim\!\bigcirc\!\sim\!\!\sim + \cdots$$

or

$$D_{\mu\nu}(k) = D^0_{\mu\nu}(k) + D^0_{\mu\mu'}(k)\Pi_{\mu'\nu'}D^0_{\nu'\nu}(k) + D^0_{\mu\mu'}\Pi_{\mu'\nu'}D^0_{\nu'\nu''}\Pi_{\nu''\mu''}D^0_{\mu''\nu} + \cdots .$$

The summation of the geometric series on the right-hand side gives

$$D_{\mu\nu}(k) = D^0_{\mu\nu}(k) + D^0_{\mu\mu'}(k)\Pi_{\mu'\nu'}D_{\nu'\nu}(k). \qquad (4.24)$$

After the substitution $D^0_{\mu\nu} = g_{\mu\nu}/k^2$, we obtain

$$k^2 D_{\mu\nu} = g_{\mu\nu} + \Pi_{\mu\nu'}D_{\nu'\nu},$$

or

$$\left[k^2 g_{\mu\nu'} - \Pi_{\mu\nu'}(k)\right]D_{\nu'\nu} = g_{\mu\nu}. \qquad (4.25)$$

The polarization operator $\Pi_{\mu\nu}$ is a second rank symmetric tensor which depends only on one vector k_μ. The most general form for such a tensor is

$$\Pi_{\mu\nu}(k) = g_{\mu\nu}a_1(k^2) + k_\mu k_\nu a_2(k^2). \qquad (4.26)$$

Similarly, the most general structure for the exact Green function $D_{\mu\nu}$ is

$$D_{\mu\nu}(k) = g_{\mu\nu}d_1(k^2) + k_\mu k_\nu d_2(k^2). \qquad (4.27)$$

The scalar functions in (4.26) and (4.27) are connected by the equation for the photon Green function (4.25). Inserting representation (4.26) into (4.25),

$$[k^2 - a_1(k^2)]D_{\mu\nu}(k) - k_\mu k_{\nu'}D_{\nu'\nu}a_2(k^2) = g_{\mu\nu},$$

and taking into account (4.27) we obtain

$$[k^2 - a_1(k^2)]d_1(k^2)g_{\mu\nu} + [k^2 - a_1(k^2)]k_\mu k_\nu d_2(k^2)$$
$$- [k_\mu k_\nu d_1(k^2)a_2(k^2) + k^2 d_2(k^2)a_2(k^2)k_\mu k_\nu] = g_{\mu\nu}. \qquad (4.28)$$

Comparing the coefficients at $g_{\mu\nu}$, we have

$$[k^2 - a_1(k^2)]d_1(k^2) = 1,$$

or

$$d_1(k^2) = \frac{1}{k^2 - a_1(k^2)}. \qquad (4.29)$$

Similarly, comparing the coefficients at $k_\mu k_\nu$, we get

$$d_2(k^2) = \frac{a_2(k^2)}{(k^2 - a_1)(k^2 - a_1 - k^2 a_2)}. \qquad (4.30)$$

Due to current conservation, the longitudinal part of $D_{\mu\nu}$, i.e. the term proportional to $k_\mu k_\nu$ in (4.27), does not contribute to the scattering amplitudes. We shall therefore concentrate on the first term in (4.27) proportional to $g_{\mu\nu}$ [§]

$$D_{\mu\nu}^t = \frac{g_{\mu\nu}}{k^2 - a_1(k^2)}. \qquad (4.31)$$

The assumption that $d_1(k^2)$ and thus $D_{\mu\nu}^t$ has a pole at $k^2 \neq 0$ would lead to a theory having nothing to do with quantum electrodynamics. The photon Green function (4.31) would have a pole at $k^2 = 0$ only if $a_1(k^2)$ vanished like k^2 at $k^2 \to 0$. Are there any theoretical reasons to expect such behaviour?

We still have not used an additional condition which current conservation imposes on the general form of the polarization operator in (4.26). Due to current conservation the amplitude M_μ of any process satisfies the condition

$$k_\mu M_\mu = 0. \qquad (4.32)$$

We have proved this relation above, for example, for the Compton effect (see (2.92)). It is valid also for processes with virtual photons, and you can check that in the proof of (2.92) for the Compton effect we did not use the

---

[§] Equation (4.27) is usually written as

$$D_{\mu\nu}(k) = g_{\mu\nu}d_1(k^2) + k_\mu k_\nu d_2(k^2) \equiv d_1(k^2)\left(g_{\mu\nu} - \frac{k_\mu k_\nu}{k^2}\right) + \tilde{d}_2(k^2)k_\mu k_\nu,$$

where

$$\tilde{d}_2 = \frac{1}{k^2} \frac{1}{k^2 - a_1 - k^2 a_2},$$

and $d_1(k^2)(g_{\mu\nu} - k_\mu k_\nu/k^2)$ is called the transverse part since $k_\mu(g_{\mu\nu} - k_\mu k_\nu/k^2) = 0$.

condition $k^2 = 0$. We will prove the hypothesis about current conservation for arbitrary processes with virtual photons below in Section 4.6.

For now, we use (4.32) for $k^2 \neq 0$. The polarization operator $\Pi_{\mu\nu}$ is an amplitude for a virtual process

$\Pi_{\mu\nu} = \sum$  ,

and, hence, it must satisfy the current conservation condition

$$k_\mu \Pi_{\mu\nu} = 0\,. \tag{4.33}$$

Inserting the general representation of the polarization operator (4.26) in (4.33), we obtain

$$k_\mu g_{\mu\nu} a_1(k^2) + k_\mu k_\mu k_\nu a_2(k^2) = 0,$$

or

$$a_1(k^2) = -k^2 a_2(k^2).$$

Then the polarization operator can be written as

$$\Pi_{\mu\nu}(k) = (g_{\mu\nu}k^2 - k_\mu k_\nu)\Pi(k^2), \qquad \Pi(k^2) = \frac{a_1(k^2)}{k^2}\,, \tag{4.34}$$

where $\Pi(k^2)$ is also often called the polarization operator.

Now we see that if $a_1(k^2) \propto k^2$ then the polarization operator $\Pi(k^2)$ (or in other words $a_2(k^2)$ in (4.26)) remains finite at $k^2 \to 0$, the total Green function $D_{\mu\nu}$ has a pole at $k^2 = 0$, the photon remains massless even after accounting for radiative corrections, and our theory is self-consistent.

Let us verify that in the framework of perturbation theory the pole of the total photon Green function does remain at $k^2 = 0$. Consider, for example, the lowest order one-loop contribution to the polarization operator

$\Pi^{(1)}_{\mu\nu} =$ 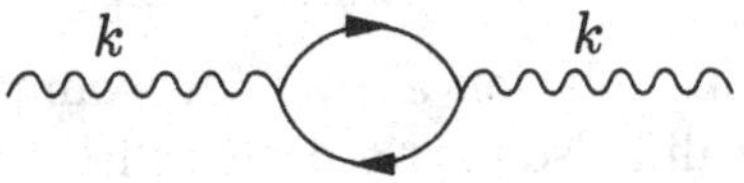

,

or, analytically,

$$\Pi^{(1)}_{\mu\nu}(k) = -e^2 \int \frac{d^4p}{(2\pi)^4 i} \,\mathrm{Tr}\left(\gamma_\mu \frac{1}{m - \hat{p}} \gamma_\nu \frac{1}{m - \hat{p} + +\hat{k}}\right).$$

(From the very structure of this expression we can already see that the polarization operator is unlikely to have a pole at $k^2 = 0$: the intermediate state contains a massive electron–positron pair, and a singularity at $k^2 = 0$ could be generated only if we had a massless intermediate state.)

In order to prove that the respective $\Pi^{(1)}(k^2)$ is finite at $k^2 \to 0$, and the photon does not acquire mass due to radiative corrections (at least in the one-loop approximation) it suffices to check that $\Pi_{\mu\nu}^{(1)}(k^2)$ is transverse, $k^\mu \Pi_{\mu\nu}^{(1)}(k^2) = 0$.

Let us calculate

$$k_\mu \Pi_{\mu\nu}^{(1)}(k) = -e^2 \int \frac{d^4p}{(2\pi)^4 i} \, \mathrm{Tr} \left(\frac{\hat{k}}{m - \hat{p}} \gamma_\nu \frac{1}{m - \hat{p} + + \hat{k}} \right). \qquad (4.35)$$

With the help of the representation $\hat{k} \equiv \hat{k} + m - \hat{p} - m + \hat{p}$ we obtain

$$k_\mu \Pi_{\mu\nu}^{(1)} = -e^2 \int \frac{d^4p}{(2\pi)^4 i} \, \mathrm{Tr} \left(-\frac{1}{m - \hat{p} + \hat{k}} \gamma_\nu + \frac{1}{m - \hat{p}} \gamma_\mu \right) = 0,$$

where we used that the integrand turns to zero after the shift of the integration variable $p - k = p'$:

$$\frac{1}{m - \hat{p}} - \frac{1}{m - \hat{p}} = 0.$$

(Note that there might be some problems with this argumentation[¶] since each integral $\propto \int d^4p/p^2$ diverges at large momenta.)

Thus we have proved that the one-loop perturbation theory contribution to the polarization operator is transverse and, hence, the photon remains massless in the one-loop approximation. The reason for the transversality of the polarization operator is current conservation which is built into quantum electrodynamics. Similar calculations can be carried out also for higher order contributions to the polarization operator, and again due to current conservation the photon remains massless in all orders of perturbation theory.

Thus, the total photon Green function is

$$D_{\mu\nu} = \frac{g_{\mu\nu}}{k^2 \left[1 - \Pi(k^2) \right]}, \qquad (4.36)$$

and due to current conservation it has a pole only at $k^2 = 0$. Like in the case of the electron Green function it is useful to write the total photon Green function in terms of the subtracted polarization operator

$$\Pi_c(k^2) = \frac{\Pi(k^2) - \Pi(0)}{1 - \Pi(0)} \qquad (4.37)$$

[¶] Still, this line of reasoning can be made more accurate by using, for example, the Pauli–Villars regularization.

in the form

$$D_{\mu\nu}(k) = \frac{Z_3 g_{\mu\nu}}{k^2[1 - \Pi_c(k^2)]} \equiv Z_3 D^c_{\mu\nu}(k^2), \qquad (4.38)$$

where $Z_3 = 1/(1 - \Pi(0))$, and we have introduced the renormalized photon Green function $D^c_{\mu\nu}(k^2)$.

Similarly to the case of the electron Green function, in calculations of scattering amplitudes with external photon lines we should take into account renormalization of the photon wave function, i.e. the factor $\sqrt{Z_3}$, which is called the photon wave function renormalization constant. Again, one factor $\sqrt{Z_3}$ for each external photon leg is associated with the amplitude, while another is compensated in the calculation of the cross section by the corresponding factor in the flux or the phase volume. Hence, effectively the scattering amplitude should be multiplied by $\sqrt{Z_3}$ for each external photon line, while normalization of the photon wave functions remains unchanged (unit residue for on-mass-shell photon).

4.4 Feynman rules for multiloop scattering amplitudes

We are now ready to give a general prescription for constructing arbitrary scattering amplitudes. Consider, for example, the process

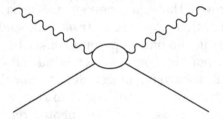

The total amplitude for this process is given by the sum of all possible multiloop diagrams with the same external lines.

We associate a free electron Green function

$$G_0 = \frac{1}{m_0 - \hat{p}} \quad \text{with each internal electron line} \quad \underrightarrow{\quad p \quad},$$

and a free photon Green function

$$D^0_{\mu\nu} = \frac{g_{\mu\nu}}{k^2} \quad \text{with each internal photon line} \quad \sim\!\!\!\sim\!\!\!\sim^{\,k}\!\!\!\sim\!\!\!\sim.$$

For the external lines

$$u(p)e^{-ipx}\sqrt{Z_2} \quad \text{corresponds to an initial electron,}$$

$$\bar{u}(p)e^{ipx}\sqrt{Z_2} \quad \text{to a final electron}$$

and, respectively,

$$\bar{v}(p)e^{-ipx}\sqrt{Z_2} \quad \text{corresponds to an initial positron,}$$

$$v(p)e^{ipx}\sqrt{Z_2} \quad \text{to a final positron.}$$

The factor

$$e_\mu^\lambda e^{-ikx}\sqrt{Z_3} \quad \text{corresponds to an initial photon}$$

and

$$e_\mu^\lambda e^{ikx}\sqrt{Z_3} \quad \text{to a final photon.}$$

Note that the external lines are on the mass shell, $p^2 = m^2$, $k^2 = 0$.

All the resulting expressions should be antisymmetrized with respect to the external electron lines, i.e.

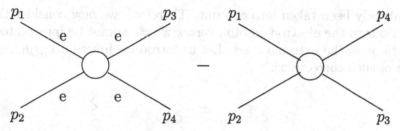

and symmetrized with respect to the final photon lines.

All internal lines in the diagrams so far are described by the bare electron and photon lines G_0 and $D_{\mu\nu}^0$. However, processes like

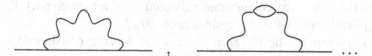

with self-energy insertions in the internal electron lines, and similar processes with polarization insertions in the internal photon lines are also possible and should be taken into account. The sums of all self-energy and polarization insertions in the internal lines give exact Green functions. We can significantly reduce the number of diagrams which we should consider, if we agree to ignore the diagrams with self-energy and polarization operator insertions in the internal lines, and instead ascribe to all internal lines not the bare but the total (exact) Green functions (4.20) and (4.38).

4.5 Renormalization of the vertex part

In the previous sections we have considered how the electron and photon propagators change when we take into account radiative corrections. Let us now turn our attention to the the photon emission amplitudes (vertex parts)

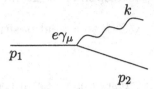

and see what happens with them if we consider higher order contributions.

Higher order processes with corrections to the electron and photon lines

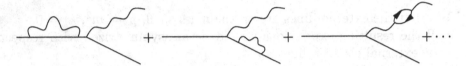

have already been taken into account. Therefore, we now consider those corrections to the electron–photon vertex which cannot be interpreted as corrections to the external lines. Let us introduce function $\Gamma_\mu(p_1, p_2)$ as a sum of such corrections:

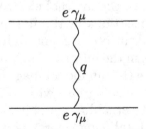

$$\Gamma_\mu(p_1, p_2) = \qquad + \qquad + \qquad + \cdots \qquad (4.39)$$

How do the radiative corrections collected in the vertex part Γ_μ change the amplitudes of the real physical processes?

Let us return to the electron–electron scattering considered earlier in Section 2.5:

At small momentum transfer q this process reduces to the usual Coulomb scattering, and from equation (2.89) we have concluded that the factor e in vertices of the tree Feynman graphs is just the charge of the electron.

However, the total vertex part is a sum of all the higher order processes which can take place in the vertex, and, hence, the charge e defined from the tree Feynman diagrams is only the first approximation to the observable charge. In order to obtain the real observable charge, all corrections to the vertex should be taken into account, and the charge should be renormalized like the mass was renormalized.

Let us represent the total vertex part $\Gamma_\mu(p_1, p_2)$ in the form

$$\Gamma_\mu(p_1, p_2) = \gamma_\mu + \Lambda_\mu(p_1, p_2), \tag{4.40}$$

where

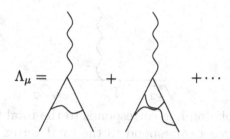

For a particle at rest and at zero momentum transfer $q = 0$,

$$\Gamma_\mu(m, m) = \gamma_\mu + \Lambda_\mu(m, m),$$

where Λ_μ is proportional to γ_μ,

$$\Lambda_\mu(m, m) = \gamma_\mu \Lambda(m, m), \tag{4.41}$$

since the matrix γ_μ is the only vector in the problem.

We define the vertex renormalization factor Z_1 by the relation

$$\Gamma_\mu(m, m) = \gamma_\mu[1 + \Lambda(m, m)] \equiv \gamma_\mu Z_1^{-1}. \tag{4.42}$$

The factor Z_1^{-1} describes how the amplitude at zero momentum transfer changes due to all possible high order processes taking place in the vertex.

The total electron–photon vertex including all radiative corrections can be written in the form

$$\begin{aligned}
\Gamma_\mu(p_1, p_2) &= \gamma_\mu + \gamma_\mu \Lambda(m, m) + \Lambda_\mu(p_1, p_2) - \Lambda_\mu(m, m) \\
&= Z_1^{-1}\left[\gamma_\mu + \frac{\Lambda_\mu(p_1, p_2) - \Lambda_\mu(m, m)}{1 + \Lambda(m, m)}\right] = Z_1^{-1}\Gamma_\mu^c,
\end{aligned}$$

where

$$\Gamma_\mu^c = \gamma_\mu + \Lambda_\mu^c \tag{4.43}$$

and

$$\Lambda_\mu^c = \frac{\Lambda_\mu(p_1, p_2) - \Lambda_\mu(m, m)}{1 + \Lambda(m, m)}. \tag{4.44}$$

Γ_μ^c (Λ_μ^c) is called the renormalized (subtracted) vertex function. In this notation it is obvious that at zero momentum transfer the total vertex function reduces to the vertex renormalization constant $\Gamma_\mu = Z_1^{-1}\gamma_\mu$.

Let us return to electron–electron scattering. The simplest one-photon exchange diagram for this process which already includes all radiative corrections to photon Green function, electron–photon vertices, and the external lines has the form

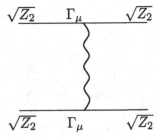

where the thick photon line corresponds to the total photon Green function, and the vertices correspond to the total vertex function Γ_μ. More complicated processes, with larger numbers of photon exchanges

Fig. 4.2

etc., also may be easily accounted for.

It is easy to see that even the most complicated diagrams contain neither bare vertices, nor bare Green functions, and only the total vertices and exact Green functions enter all expressions. There is obviously no need to consider diagrams with self-energy and polarization corrections, constructed with the use of either bare or total Green functions,

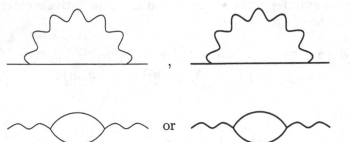

since they are already included in the Green functions. Diagrams like those in Fig. 4.2 which do not contain self-energy, polarization or vertex corrections are called skeleton diagrams.

Total vertices, electron, and photon Green functions are all multiplicatively connected with the respective renormalized functions (see (4.21), (4.38), and (4.43)), and may be written as

$$\Gamma_\mu = Z_1^{-1}\Gamma_\mu^c, \qquad (4.45)$$
$$G(p) = Z_2\,G^c(p), \qquad (4.46)$$
$$D_{\mu\nu}(k) = Z_3\,D_{\mu\nu}^c(k). \qquad (4.47)$$

Let us see what happens if we insert these representations in the diagrams. An internal electron line always starts at one vertex and ends at another, the same is true for internal photon lines. Therefore, it is convenient to write the factors Z_2 and Z_3 in (4.46), (4.47) as products of two square roots $\sqrt{Z_2}\cdot\sqrt{Z_2}$, $\sqrt{Z_3}\cdot\sqrt{Z_3}$, and associate each of these square root factors with the beginning and the end of the corresponding line. This means that, since a vertex is a point where two electron lines and one photon line meet, each vertex will be multiplied by the factor

$$e_c = eZ_1^{-1}Z_2\sqrt{Z_3}, \qquad (4.48)$$

which is called the renormalized charge. All renormalization factors Z_i disappear from the diagrams written in terms of the renormalized charge e_c and the renormalized Green functions G^c and D^c. The analytic expressions for the skeleton diagrams built up of e_c, G^c, and D^c have exactly the same form as those in terms of the bare charge and Green functions G and D.

It is easy to see that the renormalized charge is just the observable physical charge. Indeed, consider again electron scattering at small angles:

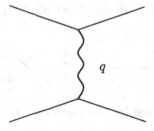

This graph has a pole at small momentum transfer $q^2 \to 0$, since

$$D^c(k) \propto \frac{1}{q^2}.$$

All other electron–electron scattering diagrams with larger numbers of photon exchanges have no poles at small momentum transfer, since they contain integrations over intermediate momenta. Hence, at small momentum transfer the main contribution to the amplitude comes from the pole diagram, and is equal to

$$A = \left(eZ_1^{-1}Z_2\sqrt{Z_3}\right)^2 \bar{u}\gamma_\mu u \frac{1}{q^2}\bar{u}\gamma_\mu u. \qquad (4.49)$$

This is just the Coulomb scattering amplitude, with the physical charge e_c, which is measured experimentally $e_c^2 \approx 4\pi/137$. We see once again that due to higher order processes not only the mass but also the electric charge of the charged particle gets renormalized; they do not coincide with the mass and the charge of the non-interacting particles.

Now consider the expression for the renormalized charge

$$e_c = eZ_1^{-1}Z_2\sqrt{Z_3}$$

from a slightly different perspective. Let us assume (in accordance with the experimental data) that different species of charged particles exist in nature, in particular, electrons (e), muons (μ) and protons (p):

$$G_e \;\rule{3cm}{0.4pt}\; e$$

$$G_\mu \;\rule{3cm}{0.4pt}\; \mu$$

$$G_p \;\rule{3cm}{0.4pt}\; p$$

Respective interaction vertices are Γ_e, Γ_μ and Γ_p. Generally speaking, they are different, since the integrals which include Green functions of particles with different masses do not necessarily lead to the same result. Let us investigate how the existence of essentially different species of particles affects the photon Green function. The exact photon Green function in a theory with only one kind of charged particles is given by the sum

If other particle species besides the electron exist, processes like

will also contribute to the photon Green function. In a sense the photon Green function is a universal function, it directly feels the presence of all species of electrically charged particles, unlike the Green functions of charged particles which are all different and depend crucially on the type of particle and its specific interactions.

4.6 The generalized Ward identity

Let us discuss an interesting theoretical problem. Imagine that the bare charges of the electron and the proton are equal. Due to interactions, observable (renormalized) charges of these particles apparently become different:

$$e_{ce} = Z_{1e}^{-1} Z_{2e} \sqrt{Z_3} \, e,$$
$$e_{cp} = Z_{1p}^{-1} Z_{2p} \sqrt{Z_3} \, e. \tag{4.50}$$

This, however, contradicts our intuitive ideas about charge conservation and would have dramatic consequences (for example, the hydrogen atom pe^- would no longer be electrically neutral). The only way to save charge conservation or, to be more precise, the universality of charge renormalization, is to assume that for each species of particle the respective Z_1 and Z_2 always coincide,

$$Z_1 = Z_2. \tag{4.51}$$

If this is true then the renormalization constants Z_1 and Z_2 which depend on the type of particle disappear from (4.50), and the physical charges of different particle species remain equal after renormalization, provided the bare charges were equal. We will prove that relation (4.51) really is valid in electrodynamics.

Let us first recall that the vertex function with on-mass-shell external fermions satisfies the equation (see Section 1.8):

$$k_\mu \Gamma_\mu(p_1, p_2) = 0, \tag{4.52}$$

where $k_\mu = p_1 - p_2$.

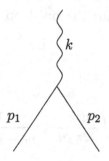

The vertex function with off-mass-shell external legs satisfies a more general condition

$$k_\mu \Gamma_\mu(p_1, p_2) = G^{-1}(p_2) - G^{-1}(p_1), \tag{4.53}$$

which is called the generalized Ward identity. We will first demonstrate that (4.53) leads to $Z_1 = Z_2$, and then we will prove the generalized Ward identity itself.

At small momentum transfer the expression on the left-hand side of (4.53) reduces to $k_\mu Z_1^{-1}\gamma_\mu$. On the right-hand side each of the Green functions gives $G_c^{-1} = m - \hat{p}$, and we obtain

$$k_\mu Z_1^{-1}\gamma_\mu = Z_2^{-1}[-\hat{p}_2 + \hat{p}_1] = Z_2^{-1}\hat{k} = Z_2^{-1}k_\mu\gamma_\mu,$$

or $Z_1 = Z_2$.

Let us now prove the generalized Ward identity (4.53). It is obviously valid for the simplest tree diagram contribution to the vertex function

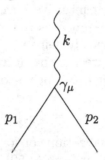

which is simply the matrix γ_μ. Indeed, the bare electron Green function is

$$G_0^{-1}(p) = m_0 - \hat{p},$$

and we have a trivial identity

$$k_\mu\gamma_\mu \equiv \hat{p}_1 - \hat{p}_2,$$

which coincides with the generalized Ward identity in the tree approximation. (For real particles $\hat{p}_1 u(p_1) = m u(p_1)$, $\bar{u}(p_2)\hat{p}_2 = \bar{u}(p_2)m$ and thus $k_\mu(\bar{u}\gamma_\mu u) = 0$.)

The analytic expression for the one-loop contribution to the vertex function has the form

$$\Lambda_\mu^{(1)} = \quad = e^2 \int \frac{d^4 k'}{(2\pi)^4 i} \gamma_\nu \frac{1}{m_0 - \hat{p}_2 + \hat{k}'} \gamma_\mu \frac{1}{m_0 - \hat{p}_1 + \hat{k}'} \gamma_\nu \frac{1}{k'^2}.$$

Calculating the contribution of this graph to $k_\mu\Gamma_\mu$, we again obtain the combination $k_\mu\gamma_\mu$ in the numerator and write it as

$$k_\mu\gamma_\mu = \hat{k} = \hat{p}_1 - \hat{p}_2 = (m_0 - \hat{p}_2 + \hat{k}') - (m_0 - \hat{p}_1 + \hat{k}'). \qquad (4.54)$$

Then

$$k_\mu \Lambda_\mu^{(1)} = e^2 \int \frac{d^4k'}{(2\pi)^4 i} \frac{1}{k'^2} \left(\gamma_\nu \frac{1}{m_0 - \hat{p}_1 + \hat{k}'} \gamma_\nu - \gamma_\nu \frac{1}{m_0 - \hat{p}_2 + \hat{k}'} \gamma_\nu \right).$$

The first term in this expression for $k_\mu \Lambda_\mu^{(1)}$ arose as the result of cancellation of the first term on the right-hand side in (4.54) with the electron propagator with momentum $p_2 - k'$. The second term is the result of a similar cancellation of the electron propagator with momentum $p_1 - k'$. Graphically the right-hand side of the expression for $k_\mu \Lambda_\mu^{(1)}$ looks as

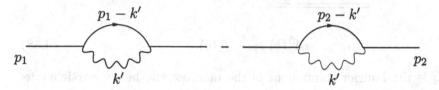

or, analytically,

$$k_\mu \Lambda_\mu^{(1)} = \Sigma^{(1)}(\hat{p}_1) - \Sigma^{(1)}(\hat{p}_2). \tag{4.55}$$

Similar expressions can be obtained for higher order diagrams. Their sum gives exactly (4.53).

The generalized Ward identity at small momentum transfer has the form

$$k_\mu \Gamma_\mu(p_1, p_2) = G^{-1}(\hat{p}_2) - G^{-1}(\hat{p}_2 + \hat{k}) = -\frac{\partial G^{-1}(p_2)}{\partial \hat{p}_2} \hat{k} \simeq -\frac{\partial G^{-1}(p)}{\partial \hat{p}} \gamma_\mu k_\mu,$$

and at zero momentum transfer degenerates into the equation

$$\Gamma_\mu(p, p) = -\frac{\partial G^{-1}(p)}{\partial \hat{p}} \gamma_\mu,$$

which may be written as

$$\Gamma_\mu(p, p) = -\frac{\partial G^{-1}(p)}{\partial p_\mu}, \tag{4.56}$$

since $\gamma_\mu = d\hat{p}/dp_\mu$.

With the proof of the generalized Ward identity we conclude our construction of quantum electrodynamics which contains only renormalized charge, renormalized vertex function, and renormalized Green functions. Further problems are connected with the study of the exact Green functions D^c and G^c, and the vertex part Γ_μ^c. In lower order approximations, however, the corresponding calculations are straightforward and relatively simple.

4.7 Radiative corrections to electron scattering in an external field

Let us consider radiative corrections to electron scattering in an external field. As we discussed in Section 2.8, such a process is just a scattering off a heavy particle. The amplitude of this process in the tree approximation has the form

$$ = e\bar{u}(p_2)\gamma_\mu u(p_1)A_\mu^0(q), \qquad (4.57)$$

p_1 _____ p_2

q

where

$$A_\mu^0(q) = \frac{e}{q^2}\mathcal{J}_\mu(q), \qquad (4.58)$$

and $\mathcal{J}_\mu$ is the Fourier component of the macroscopic heavy particle current.

What happens with this one-photon exchange diagram when we take into account processes of higher order? First, the vertex gets dressed, and an additional factor $\sqrt{Z_2}$ arises for each electron line:

$\sqrt{Z_2} \quad \Gamma_\mu \quad \sqrt{Z_2}$

$$ = e\bar{u}(p_2)\Gamma_\mu^c(p_2, p_1)u(p_1)A_\mu(q), \qquad (4.59)$$

where $\Gamma_\mu = Z_1^{-1}\Gamma_\mu^c$. Second, the external field A_μ also changes. Indeed, according to (4.58), it has the form

$$A_\mu^0(q) = eD_{\mu\nu}^0(q)\mathcal{J}_\nu(q),$$

and the higher order corrections lead to

$$A_\mu(q) = \frac{eZ_3}{q^2[1 - \Pi^c(q^2)]}\mathcal{J}_\mu(q). \qquad (4.60)$$

The field is modified due to various processes with virtual electron–positron pairs, such as

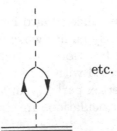

etc.

The final expression for the scattering amplitude is

$$F = e_c \bar{u}(p_2)\Gamma^c_\mu(p_2, p_1)u(p_1)\frac{1}{1 - \Pi^c(q^2)}A^0_\mu(q) \qquad (4.61)$$

with $A^0_\mu(q)$ defined in (4.58).

Thus the amplitude is modified due to two effects, namely, renormalization of the interaction, and change of the external field. The external field changes as a result of production of virtual electron–positron pairs in the vacuum. The sign of this change may be easily determined from physical considerations. Assume, for example, that the heavy particle has a positive charge. If an electron–positron pair is created in the field of this particle, the positron is repulsed and moves away to a larger distance:

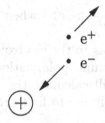

The heavy particle is thus surrounded by negative charges, and its observable charge decreases. This effect is called vacuum polarization, and it is similar to the polarization of a dielectric by an external free charge. In the case of a dielectric the molecular dipoles play the rôle of the electron–positron pairs.

The physical and bare charges are connected by the relationship (4.48) (recall that $Z_1 = Z_2$ due to the Ward identity)

$$e_c^2 = Z_3 e_0^2.$$

Our simple consideration immediately leads to the conclusion that $Z_3 < 1$, and the physical charge is screened. It is just this screened charge squared

$$e_c^2 = \frac{4\pi}{137},$$

that we observe at macroscopic distances. If we came closer to the electron, we would start feeling its bare charge squared which is larger than $4\pi/137$.

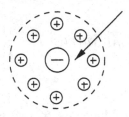

Thus we can expect that

$$\Pi^c(q^2) > 0, \quad \Pi^c(0) = 0,$$

and the interaction would grow at large momentum transfers (corresponding to small distances):

$$e_0^2 \sim \frac{e_c^2}{1 - \Pi^c(q^2)} > e_c^2 \quad \text{when} \quad |q^2| \gg m^2.$$

As we have seen, corrections to the electron scattering amplitude in an external field are due to vacuum polarization and to corrections to the vertex function. Now we will calculate the contributions of both these effects to the scattering amplitude to first order in e^2.

4.7.1 One-loop polarization operator

According to the Feynman rules the electron–positron contribution to the polarization operator

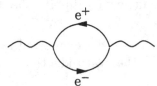

has the form

$$\Pi_{\mu\nu}(k) = -e^2 \int \frac{d^4p}{(2\pi)^4 i} \, \text{Tr} \left(\gamma_\mu \frac{1}{m - \hat{p}} \gamma_\nu \frac{1}{m - \hat{p} + \hat{k}} \right). \tag{4.62}$$

Other particles also contribute to the vacuum polarization operator. The muon contribution can be calculated similarly to (4.62). Calculation of the proton–antiproton contribution

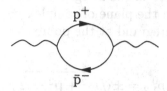

is much harder, since besides the electromagnetic interaction the protons are also subject to strong interaction. This is still an unsolved problem but, as we will see, at small momenta the contribution to the polarization operator of any particle–antiparticle pair behaves like $\sim k^2/m^2$, where m is the mass of the respective particles. Thus, at small energies even the muon contribution is negligible. More careful investigations of radiative corrections at not too high energies guarantee that there are no unknown light charged particles.

Returning to the calculation of the polarization operator $\Pi_{\mu\nu}$, let us write it in the form

$$\Pi_{\mu\nu} = (g_{\mu\nu}k^2 - k_\mu k_\nu)\Pi(k^2).$$

Then

$$\Pi_{\mu\mu} = 3k^2\Pi(k^2) = -e^2 \int \frac{d^4p}{(2\pi)^4 i} \frac{\mathrm{Tr}\left[\gamma_\nu(m+\hat{p})\gamma_\nu(m+\hat{p}-\hat{k})\right]}{(m^2-p^2)(m^2-(p-k)^2)}. \quad (4.63)$$

The trace in the numerator is easy to calculate:

$$\mathrm{Tr}\left[\gamma_\nu(m+\hat{p})\gamma_\nu(m+\hat{p}-\hat{k})\right] = \mathrm{Tr}\left[(4m-2\hat{p})(m+\hat{p}-\hat{k})\right]$$
$$= 16m^2 - 8p(p-k). \quad (4.64)$$

In this calculation we made use of the fact that the trace of the product of an odd number of γ-matrices is zero, and also used the auxiliary relations for γ-matrices

$$\gamma_\nu\gamma_\nu = 4, \quad \gamma_\nu\hat{p}\gamma_\nu = -2\hat{p}, \quad \mathrm{Tr}\,\gamma_\mu\gamma_\nu = 4g_{\mu\nu}.$$

Substituting (4.64) in (4.63) we see that the integral (4.63) diverges at large integration momenta. However, radiative corrections are determined by the subtracted polarization operator $\Pi^c(k^2)$,

$$\Pi^c(k^2) = \frac{\Pi(k^2) - \Pi(0)}{1 - \Pi(0)} \propto \Pi(k^2) - \Pi(0), \quad (4.65)$$

which is convergent.[||]

We start the calculation of the integral (4.63) using the analytic properties of the integrand in the plane of complex p_0. The first denominator generates poles of the integrand at the points

$$p_{1,2}^0 = \pm\sqrt{m^2 + \mathbf{p}^2 - i\varepsilon}\,,$$

the second one generates poles at the points

$$p_{3,4}^0 = k_0 \pm \sqrt{m^2 + (\mathbf{p} - \mathbf{k})^2 - i\varepsilon}\,.$$

We will calculate the polarization operator at space-like momenta $k^2 < 0$, its values at time-like momenta will be obtained by analytic continuation. For space-like k we can always choose a reference frame where $k_0 = 0$. In this reference frame the poles are symmetric with respect to the imaginary axis in the energy plane as shown in Fig. 4.3.

Then we can rotate the integration contour from the real to the imaginary axis. This is allowed since in the process of rotation the contour does not cross any singularities.

[||] Rigorously speaking, the integral (4.63) is quadratically divergent and one subtraction in (4.65) would not make it convergent. However, we have to recall that the polarization operator $\Pi(k^2)$ was essentially defined as the factor before the tensor $k_\mu k_\nu$ in (4.26) in the representation of the polarization operator in terms of independent tensors. Comparing representation (4.26) with the explicit integral for the one-loop contribution to the polarization operator $\Pi_{\mu\nu}(k^2)$ in (4.62) it is easy to see that the integral for $\Pi(k^2)$ defined in this way diverges only logarithmically. Then one subtraction in (4.65) really makes it convergent. We have introduced a fake quadratic divergence in the integral (4.63) when we carelessly assumed that the integral in (4.62) has the proper transverse structure. Sure, such structure is imposed by current conservation, but literally we are dealing with the divergent integral, and this condition may be apparently violated by the divergent terms. Anyway, it is easy to check that the fake quadratically divergent term in (4.63) is real, and thus our careless treatment of the integral (4.63) did not change either the analytic structure or the imaginary part of this integral. We are going to calculate the integral (4.63) using the analytic properties of the integrand and, hence, we can ignore all complications mentioned here.

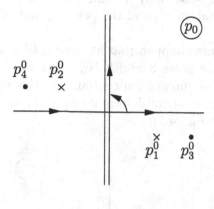

Fig. 4.3

The integral after the rotation is obviously real because only quadratic expressions in p_0 enter the integrand in (4.63), and the integration volume element

$$\frac{d^4 p}{(2\pi)^4 i} = \frac{dp_0 d^3 p}{(2\pi)^4 i} = \frac{dp_0' d^3 p}{(2\pi)^4}$$

is also real after the rotation. Hence, the integral (4.63) is real on the real negative half-axis (bold line in Fig. 4.4) in the complex k^2 plane.

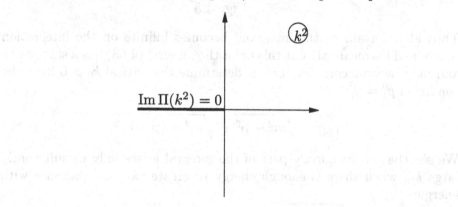

Fig. 4.4

Moreover, the polarization operator does not have any singularities at negative k^2, since the denominator in (4.63) does not turn to zero anywhere on the rotated contour. (To secure convergence of the integral at large p^2, we can cut off large integration momenta by an auxiliary parameter Λ. We will return to this problem later.) Let us see where in the

k^2 plane the integral (4.63) may become complex. Positions of the poles in the p_0 plane depend on k^2, and the poles p_3^0 and p_4^0 move as functions of k.

For time-like k it may happen that at certain k^2 either the poles 4 and 1 (for $k_0 > 0$) or the poles 3 and 2 ($k_0 < 0$) will collide ($p_4^0 = p_1^0$ or $p_3^0 = p_2^0$) and pinch the integration contour. At this moment the contour is immobilized, and one cannot deform it to avoid zero in the denominator of the integrand (see Fig. 4.5).

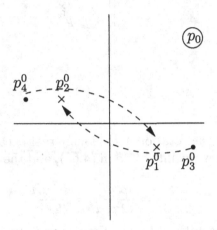

Fig. 4.5

Thus at a certain k_0 the integrand becomes infinite on the integration contour. This means that at this point the integral (4.63) has a singularity and may become complex. Let us determine the critical $k_0 > 0$ from the condition $p_4^0 = p_1^0$:

$$k_0 = \sqrt{m^2 + \mathbf{p}^2} + \sqrt{m^2 + (\mathbf{p} - \mathbf{k})^2}. \qquad (4.66)$$

We see that an imaginary part of the integral arises only at sufficiently large k_0, when there is enough energy to create two real particles with energies

$$\sqrt{m^2 + \mathbf{p}^2} \qquad \text{and} \qquad \sqrt{m^2 + (\mathbf{p} - \mathbf{k})^2}$$

(The integral can also acquire an imaginary part at

$$k_0 = -\sqrt{m^2 + \mathbf{p}^2} - \sqrt{m^2 + (\mathbf{p} - \mathbf{k})^2}, \qquad (4.67)$$

corresponding to the same point in k^2. Physically this case is not interesting since it corresponds to negative photon energy.)

Apparently, the condition (4.66) for the position of singularity is not relativistically invariant. To write it in a relativistically invariant way, let us go to the reference frame where $\mathbf{k} = 0$ (this can be done because $k^2 > 0$). Then

$$k_0^2 = k^2 = 4(m^2 + \mathbf{p}^2),$$

and the singularity arises at

$$k^2 \geq 4m^2. \tag{4.68}$$

Consider the k^2 plane cut from the point $4m^2$ along the real axis to infinity:

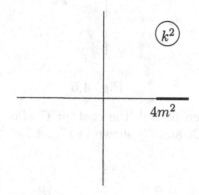

It is clear that the integral (4.63) has no other singularities in the complex k^2 plane besides the singularities on the positive real axis. Then it is an analytic function in the cut k^2 plane, and we can write a dispersion representation for $\Pi^c(k^2)$:

$$\Pi^c(k^2) = \frac{1}{\pi} \int_{4m^2}^{\infty} \frac{dk'^2 \operatorname{Im} \Pi^c(k'^2)}{k'^2 - k^2}. \tag{4.69}$$

If this integral is divergent at large k'^2 (which it is), we can improve its convergence by subtracting $\Pi^c(0)$ since the physical polarization operator should satisfy the condition $\Pi^c(0) = 0$ anyway. Then we obtain

$$\Pi^c(k^2) = \frac{1}{\pi} \int dk'^2 \operatorname{Im} \Pi^c(k'^2) \left[\frac{1}{k'^2 - k^2} - \frac{1}{k'^2} \right],$$

and, since an imaginary part exists only at $k^2 > 4m^2$,

$$\Pi^c(k^2) = \frac{k^2}{\pi} \int_{4m^2}^{\infty} \frac{dk'^2 \operatorname{Im} \Pi^c(k'^2)}{k'^2(k'^2 - k^2)}. \tag{4.70}$$

Let us now calculate $\text{Im}\,\Pi^c$. The pole p_4^0 moves with k_0 increasing and drags the integration contour, and the imaginary part of $\Pi_{\mu\mu} = 3k^2\Pi(k^2)$ arises when the contour C passes the pole p_1^0 (see Fig. 4.6).

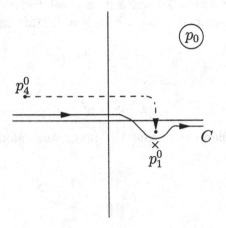

Fig. 4.6

After the pole has been passed, the contour C may be represented as a sum of two contours C_1 and C_2 shown in Fig. 4.7.

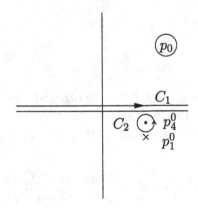

Fig. 4.7

The contour C_1 is a straight line slightly above the real axis, while the contour C_2 is a closed loop around the pole at p_4^0. It is easy to see that the integral over C_1 is real, and an imaginary part is due to integration over C_2. This last integral is just the residue at the point $p_0 = p_4^0$, and we obtain the pole contribution to $3k^2\Pi(k^2)$

$$-e^2 \int \frac{d^3p}{(2\pi)^3} \frac{\text{Tr}\,(\ldots)}{(m^2 - p^2 - i\varepsilon)} \frac{1}{2(k_0 - \sqrt{m^2 + (\mathbf{p} - \mathbf{k})^2})}.$$

Then the calculation of the imaginary part of the polarization operator,

$$3k^2 \operatorname{Im} \Pi(k^2) = -e^2 \pi \int \frac{d^3 p}{(2\pi)^3} \frac{\delta(m^2 - p^2)}{2(k_0 - \sqrt{m^2 + (\mathbf{p} - \mathbf{k})^2})} \operatorname{Tr}(\ldots),$$

is easy using (1.164):

$$\operatorname{Im} \frac{1}{m^2 - p^2 - i\varepsilon} = \pi \delta(m^2 - p^2) = \pi \delta_+(m^2 - p^2) \quad \text{(since } p_0 = p_4^0 > 0).$$

Writing the integration volume as

$$\frac{d^3 p}{2(k_0 - \sqrt{m^2 + (\mathbf{p} - \mathbf{k})^2})} = d^4 p\, \theta\left(k_0 - \sqrt{m^2 + (\mathbf{p} - \mathbf{k})^2}\right) \delta\left(m^2 - (p - k)^2\right)$$

$$\equiv d^4 p\, \delta_+\left(m^2 - (p - k)^2\right),$$

we immediately come to an explicitly relativistically invariant representation for the imaginary part of the polarization operator:

$$3k^2 \operatorname{Im} \Pi(k^2)$$

$$= -e^2 \pi \int \frac{d^4 p}{(2\pi)^3} \delta_+\left(m^2 - p^2\right) \delta_+\left(m^2 - (p - k)^2\right) \operatorname{Tr}(\ldots). \tag{4.71}$$

This imaginary part coincides with what we would obtain from the unitarity condition (compare with the discussion in 3.3.4), i.e. with the expression given by the diagram

containing real particles (see the rules for calculation of the diagrams with real internal particles in Section 1.13). Thus, we have confirmed once again that the Feynman diagrams satisfy the unitarity condition automatically.

Using the explicit form (4.64) for $\operatorname{Tr}(\ldots)$ we derive from (4.71)

$$3k^2 \operatorname{Im} \Pi(k^2)$$

$$= -e^2 \int \frac{d^4 p}{8\pi^2} [16m^2 - 8p^2 + 8pk] \delta_+(m^2 - p^2) \delta_+(m^2 - (p - k)^2). \tag{4.72}$$

Due to the condition

$$p^2 = m^2 \quad \Longrightarrow \quad m^2 - (p - k)^2 = 2pk - k^2 = 0 \quad \Longrightarrow \quad 2pk = k^2$$

imposed by the δ-functions, the integral in equation (4.72) reduces to

$$\operatorname{Im}\Pi(k^2) = -e^2 \frac{8m^2 + 4k^2}{3k^2} \int \frac{d^4p}{8\pi^2} \delta_+(p^2 - m^2)\,\delta_+((p-k)^2 - m^2).$$

The simplest way to calculate this integral is to go, once again, to the reference frame where $\mathbf{k} = 0$. Then $(p - k)^2 - m^2 = -2p_0 k_0 + k^2$ and

$$\int \frac{d^4p}{8\pi^2}\delta_+(p^2 - m^2)\delta_+(k^2 - 2p_0 k_0) = \int \frac{d^3p}{8\pi^2}\delta(p_0^2 - p_c^2 - m^2)\frac{1}{2k_0}, \quad (4.73)$$

where $p_0 = k^2/2k_0 = k_0/2$ and $p_c \equiv |\mathbf{p}|$. In spherical coordinates, $d^3p = p_c^2 dp_c \times 4\pi = 2\pi p_c dp_c^2$, and the integrand in (4.73) can be written as

$$\frac{1}{8\pi^2}\frac{1}{2k_0}2\pi\sqrt{p_0^2 - m^2} = \frac{1}{8\pi}\frac{1}{k_0}\frac{\sqrt{k_0^2 - 4m^2}}{2} = \frac{1}{16\pi}\sqrt{\frac{k^2 - 4m^2}{k^2}}.$$

Then

$$\begin{aligned}
\operatorname{Im}\Pi(k^2) &= -\frac{e^2}{16\pi}\frac{8m^2 + 4k^2}{3k^2}\sqrt{\frac{k^2 - 4m^2}{k^2}} \\
&= -\frac{e^2}{4\pi}\frac{2m^2 + k^2}{3k^2}\sqrt{\frac{k^2 - 4m^2}{k^2}},
\end{aligned}$$

and finally

$$\operatorname{Im}\Pi(k^2) = -\frac{\alpha}{3}\left(1 + \frac{2m^2}{k^2}\right)\sqrt{1 - \frac{4m^2}{k^2}}. \quad (4.74)$$

Let us note that due to (4.65), $\operatorname{Im}\Pi(k^2)$ equals $\operatorname{Im}\Pi^c(k^2)$. Hence,

$$\Pi^c(k^2) = -\frac{\alpha k^2}{3\pi}\int_{4m^2}^{\infty}\frac{d\kappa^2\sqrt{1 - \frac{4m^2}{\kappa^2}}}{\kappa^2(\kappa^2 - k^2)}\left(1 + \frac{2m^2}{\kappa^2}\right). \quad (4.75)$$

Let us calculate the polarization operator (4.75) in two special cases of small and large virtualities k^2.

For $k^2 \to 0$, we obtain after the substitution $x = 4m/\kappa^2$

$$\Pi^c(k^2) \simeq -\frac{\alpha k^2}{3\pi 4m^2}\int_0^1 dx\sqrt{1 - x}\left(1 + \frac{x}{2}\right) = -\frac{\alpha k^2}{15\pi m^2}, \quad (4.76)$$

i.e. the polarization operator vanishes as $\Pi^c(k^2) \propto k^2$ when $k^2 \to 0$, as expected.

For large negative virtualities, $|k^2| \to \infty$, the main, logarithmically growing, contribution to the integral comes from the integration region with momenta $4m^2 \ll \kappa^2 \ll |k^2|$, where the integrand behaves as $d\kappa^2/\kappa^2$. This leading contribution can be easily calculated:

$$\Pi^c(k^2) \simeq \frac{\alpha}{3\pi} \int_{4m^2}^{-k^2} \frac{d\kappa^2}{\kappa^2} \simeq \frac{\alpha}{3\pi} \ln \frac{-k^2}{m^2}. \tag{4.77}$$

The logarithm in (4.77) is in fact the main problem of quantum electrodynamics. For example, this logarithm enters the amplitude for electron scattering by an external field with momentum transfer q ($q^2 < 0$) through the factor

$$\frac{1}{1 - \Pi^c(q^2)} = \left(1 - \frac{\alpha}{3\pi} \ln \frac{-q^2}{m^2}\right)^{-1},$$

which means that the amplitude acquires a pole at very large space-like momentum transfer. This singularity has an obscure physical meaning, and this is a real difficulty. We will temporarily postpone consideration of this problem and turn instead to the radiative corrections to the interaction vertex.

4.7.2 One-loop vertex part

Let us look at the scattering amplitude given in (4.61),

$$F = e_c \bar{u}(p_2) \Gamma^c_\mu(p_2, p_1) u(p_1) \frac{1}{1 - \Pi^c(q^2)} A^0_\mu(q),$$

from a new perspective. For small momentum transfer q^2

$$\frac{1}{1 - \Pi^c(q^2)} \simeq 1 + \Pi^c(q^2),$$

and taking into account that

$$\Gamma^c_\mu = \gamma_\mu + \Lambda^c_\mu,$$

we obtain

$$F \simeq e_c \bar{u}(p_2) \left[\gamma_\mu (1 + \Pi^c(q^2)) + \Lambda^c_\mu(p_2, p_1) \right] u(p_1) A^0_\mu(q). \tag{4.78}$$

All terms in the square brackets here are proportional to γ_μ, and the factor before γ_μ differs from unity due to contributions of the vacuum polarization and the vertex part. The non-vanishing vertex part contribution Λ^c_μ demonstrates that the charge distribution inside the electron has a finite radius. Indeed, the usual quantum mechanical form factor for a particle of finite size, at small momenta transfer has the form

$$F(q) \simeq 1 + \frac{q^2 r_0^2}{6},$$

where r_0 is the mean radius of the charge distribution.

In the first order of perturbation theory the vertex part Λ_μ is

$$\Lambda_\mu = \quad = e_c^2 \int \frac{d^4 k}{(2\pi)^4 i} \gamma_\nu \frac{1}{m - \hat{p}_2 + \hat{k}} \gamma_\mu \frac{1}{m - \hat{p}_1 + \hat{k}} \gamma_\nu \frac{1}{k^2}, \quad (4.79)$$

and

$$\Lambda_\mu^c = \Lambda_\mu - \Lambda_\mu(m, m). \quad (4.80)$$

In Section 2.4 we have written the most general expression for the electron–photon vertex on the mass shell:

$$\bar{u}(p_2)\Lambda_\mu^c(p_2, p_1)u(p_1) = \bar{u}(p_2)\left[a\gamma_\mu + b\,\sigma_{\mu\nu}q_\nu \right]u(p_1). \quad (4.81)$$

The total vertex part was parametrized as

$$\Gamma_\mu = \tilde{a}\gamma_\mu + b\,\sigma_{\mu\nu}q_\nu,$$

where $\tilde{a} = \tilde{a}(q^2)$, $b = b(q^2)$ and $\tilde{a}(0)$ is simply the charge of the particle. According to (4.80), the function $a(q^2)$ vanishes at zero q^2

$$a(0) = 0,$$

while $b(0)$ may be different from zero. We have shown in Section 2.4 that if $b = 0$, the electron has a magnetic moment equal to the Bohr magneton. The non-vanishing $b \neq 0$ means that due to interactions, the electron acquires an additional magnetic moment which is called the anomalous magnetic moment. It is clear from the explicit form of (4.79) that there are no special reasons to expect $b(0) = 0$. Direct calculations confirm that $b(0) \neq 0$.

We have constructed electrodynamics starting with the simplest assumption that the electron–photon interaction is described by a vertex with $\tilde{a}(q^2) = \text{const} = e$ and $b(q^2) = 0$. However, a non-trivial electric form factor (dependence of $\tilde{a}$ on q^2) and an anomalous magnetic moment arise when higher order contributions are taken into account. This effect can be easily understood qualitatively. Let us consider the process

Even if the electron was initially at rest, it acquires a non-vanishing momentum after the emission of a virtual photon. Hence, the external field interacts with a current

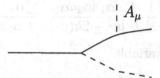

and not with a static charge. Electron motion, naturally, generates a magnetic moment. An electric form factor arises since the charge in this process is effectively distributed over a finite region $r_0 \sim 1/m$:

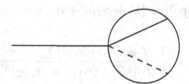

First order perturbation theory correction to the electron magnetic moment was calculated by Schwinger [3] in 1948:

$$\mu = \mu_0 \left(1 + \frac{\alpha}{2\pi} \right). \tag{4.82}$$

We will briefly outline the calculation of the one-loop vertex part $\Lambda_\mu^{(1)}$ by the Feynman method. The explicit expression for Λ_μ in the first order in e_c^2 is

$$\Lambda_\mu^{(1)} = \quad = e_c^2 \int \frac{d^4k}{(2\pi)^4 i} \frac{\gamma_\nu (m + \hat{p}_2 - \hat{k})\gamma_\mu (m + \hat{p}_1 - \hat{k})\gamma_\nu}{[m^2 - (p_2 - k)^2][m^2 - (p_1 - k)^2]k^2}. \tag{4.83}$$

To calculate such integrals Feynman invented the identity

$$\frac{1}{abc} = \frac{1}{3!} \int_0^1 \frac{d\alpha_1 d\alpha_2 d\alpha \delta(\alpha_1 + \alpha_2 + \alpha_3 - 1)}{(a\alpha_1 + b\alpha_2 + c\alpha_3)^3}. \tag{4.84}$$

In our case

$$\frac{1}{abc} = \frac{1}{[m^2 - (p_2 - k)^2][m^2 - (p_1 - k)^2]k^2}$$

$$= \frac{1}{3!} \int \frac{d\alpha_1 d\alpha_2 d\alpha_3 \, \delta(\sum \alpha_i - 1)}{\{\alpha_3 k^2 + \alpha_2[(p_2 - k)^2 - m^2] + \alpha_1[(p_1 - k)^2 - m^2]\}^3}.$$

The δ-function gives $\alpha_3 = 1 - \alpha_1 - \alpha_2$, and then

$$\frac{1}{abc} = \frac{1}{3!} \int \frac{d\alpha_1 d\alpha_2 d\alpha_3 \delta(\sum \alpha_i - 1)}{[k^2 - 2k(\alpha_1 p_1 + \alpha_2 p_2)]^3}.$$

Shifting the integration variable

$$k' = k - \alpha_1 p_1 - \alpha_2 p_2,$$

we get rid of the term linear in the integration momentum in the integrand,

$$k[k - 2(\alpha_1 p_1 + \alpha_2 p_2)] = (k' + \alpha_1 p_1 + \alpha_2 p_2)[k' - (\alpha_1 p_1 + \alpha_2 p_2)],$$

and obtain

$$\frac{1}{abc} = \frac{1}{3!} \int \frac{d\alpha_1 d\alpha_2 d\alpha_3 \delta(\sum \alpha_i - 1)}{[k'^2 - (\alpha_1 p_1 + \alpha_2 p_2)^2]^3}. \tag{4.85}$$

Then

$$\Lambda_\mu = \frac{1}{3!} \int d\alpha_1 d\alpha_2 d\alpha_3 \delta(\sum \alpha_i - 1) e_c^2 \int \frac{d^4 k'}{(2\pi)^4 i}$$

$$\times \frac{\gamma_\nu[m + (1 - \alpha_2)\hat{p}_2 - \alpha_1 \hat{p}_1 - \hat{k}']\gamma_\mu[m + (1 - \alpha_1)\hat{p}_1 - \alpha_2 \hat{p}_2 - \hat{k}']\gamma_\nu}{[k'^2 - (\alpha_1 + \alpha_2)^2 m^2 + \alpha_1 \alpha_2 q^2]^3},$$

where we have taken into account that $q^2 = (p_1 - p_2)^2 = 2m^2 - 2p_1 p_2$. We now omit the terms linear in k' in the numerator of the integrand since they obviously give no contribution to the integral. The remaining terms in the numerator have the form

$$f_1(q^2, \alpha_1, \alpha_2)\gamma_\mu + f_2(q^2, \alpha_1, \alpha_2)\sigma_{\mu\nu}q_\nu + k'^2 \gamma_\mu. \tag{4.86}$$

Thus, the calculation of the momentum integral reduces to the calculation of two standard integrals:

$$I_1 = \int \frac{d^4 k'}{(2\pi)^4 i} \frac{1}{(k'^2 - \Delta)^3} \tag{4.87}$$

and

$$I_2 = \int \frac{d^4k'}{(2\pi)^4 i} \frac{k'^2}{(k'^2 - \Delta)^3}. \tag{4.88}$$

In these integrals we can rotate the integration contour in the complex k_0' plane from the real axis to the imaginary axis because the contour does not cross any physical singularities in the process of rotation.

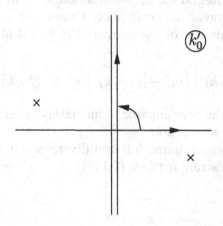

The rotation effectively reduces to the change of variables

$$k_0' = i k_4' \quad \text{and} \quad k'^2 = k_0'^2 - \mathbf{k}'^2 = -k_4'^2 - \mathbf{k}'^2,$$

and after the rotation, integration effectively goes over the four-dimensional Euclidean space. Angular integration in spherical coordinates is trivial,

$$d^4k' = \pi^2 k'^2 dk'^2,$$

and we obtain

$$I_1 = -\int \frac{d^4k'}{(2\pi)^4} \frac{1}{(k'^2 + \Delta)^3} = -\int \frac{\pi^2 k'^2 dk'^2}{(2\pi)^4 (k'^2 + \Delta)^3}$$

$$= -\int_0^\infty \frac{\pi^2 x \, dx}{(2\pi)^4 (x + \Delta)^3} = -\frac{1}{16\pi^2 \Delta}.$$

Similarly,

$$I_2 = \int \frac{\pi^2 k'^4 dk'^2}{(2\pi)^4 (k'^2 + \Delta)^3} = \int_0^\infty \frac{\pi^2 x^2 dx}{(2\pi)^4 (x + \Delta)^3}$$

$$= \frac{1}{16\pi^2} \int_\Delta^\infty \frac{(y - \Delta)^2 dy}{y^3} = \frac{1}{16\pi^2} \left[\int_\Delta^\infty \frac{dy}{y} - 2\Delta \int_\Delta^\infty \frac{dy}{y^2} + \Delta^2 \int_\Delta^\infty \frac{dy}{y^3} \right]$$

$$= \frac{1}{16\pi^2} \left[\ln \left(\frac{y}{\Delta} \right) - \frac{3}{2} \right].$$

The last integral is logarithmically divergent at large integration momenta. This ultraviolet divergence in the vertex function can be removed by renormalization, i.e. by subtracting $\Lambda_\mu(m, m)$ which contains exactly the same divergent logarithm $\ln(y/\Delta(q = 0))$.

However, the subtraction creates a new problem. The electron momenta in the subtraction term are on the mass shell ($p_1^2 = m^2$ and $p_2^2 = m^2$), and the integral for the subtraction term in this case becomes logarithmically divergent at *small* photon momenta k. Indeed, for real electrons the denominator of the integrand in (4.83) at small k^2 has the form

$$\left[m^2 - (p_2 - k)^2\right]\left[m^2 - (p_1 - k)^2\right]k^2 \simeq (2p_2k)(2p_1k)\,k^2.$$

High power of k in the denominator immediately generates a logarithmic divergence $\int d^4k/k^4$ at small k.

We have already encountered infrared divergence in Section 2.9, when we considered bremsstrahlung (see (2.148))

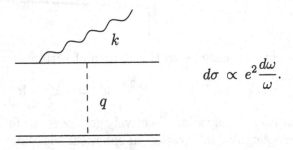

$$d\sigma \propto e^2 \frac{d\omega}{\omega}.$$

As mentioned there, the infrared divergence is the result of incorrect treatment of the scattering problem. Indeed, for sufficiently small ω the expansion parameter is not small, and the perturbation theory cannot be applied. On the other hand, the problem is unphysical in the following sense: in any experiment, as soon as a charged particle is born, photons are also created. There is no way to create a charged particle without accompanying photons, since the particle always emits soft photons under the influence of an arbitrary small perturbation (and the smaller the photon frequency, the larger their number). How can one overcome this difficulty? We could assume that the initial state consists of an electron and a large number of photons, i.e.

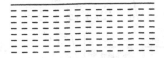

However, such an approach would suffer from a certain ambiguity. The bulk matter is always neutral (atoms are neutral), and the number of photons in the initial state would depend on how the charged particle was produced.

Thus, the only consistent way to treat the problem is to start with neutral matter and to take into account the real production process of charged particles, for example

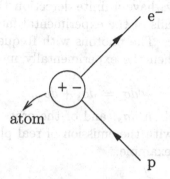

From the physical considerations above it is clear that the probability to create an electron without accompanying photons is zero. However, a perturbative calculation of the cross section of, for instance, the process

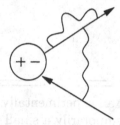

generates infinitely large corrections.

In fact, we also have to consider processes with production of many photons. Suppose, for instance, that two non-relativistic particles of opposite charges e and $-e$ and energies $\sim \epsilon$ are produced with relative velocity v. Then the cross section of a process in which n accompanying photons are emitted has the structure

$$\sigma_n \sim \frac{1}{n!} \left(\frac{\alpha}{\pi} v^2 \ln \frac{\epsilon}{\omega} \right)^n \exp \left\{ -\frac{\alpha}{\pi} v^2 \ln \frac{\epsilon}{\omega} \right\}.$$

We see that the probability of emission of any fixed number of photons tends to zero with $\omega \to 0$. Summation over all possible emissions leads, however, to a constant cross section,

$$\sum_n \sigma_n = \text{const.}$$

To discuss the probability of emitting exactly n very soft photons would be meaningful only if we had the experimental means to detect photons with fantastically small frequencies. Experimentally the parameter $\alpha/\pi \sim 1/500$ is very small, hence, the infrared logarithm becomes large (and the respective radiative effects measurable) only at academically small frequencies.

In practice, a different approach to scattering processes is used. Real photon detectors always have a finite detection threshold $\omega_{\min}$, which is determined by the details of the experimental facility, by the sensitivity of the instruments, etc. The photons with frequencies $\omega < \omega_{\min}$ always escape undetected. Then the experimentally measured cross section is the sum

$$d\sigma = d\sigma_s + d\sigma_\gamma,$$

of the elastic cross section $d\sigma_s$, and of the cross section $d\sigma_\gamma$ that sums all inelastic processes with the emission of real photons with frequencies smaller than $\omega_{\min}$, for example,

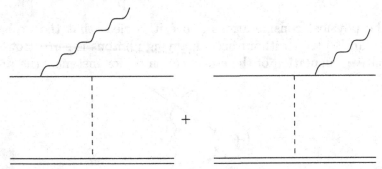

In order to calculate such an experimentally measured cross section it is convenient to introduce temporarily a small photon mass λ. Then the free photon Green function is proportional to

$$\frac{1}{k^2 - \lambda^2},$$

and calculating the vertex part Λ_μ

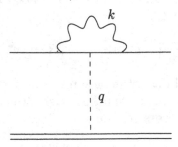

at small momenta transfer $|q^2|/m^2 \ll 1$, we obtain

$$\Lambda_\mu^c = \gamma_\mu \frac{\alpha}{3\pi} \left[\ln \frac{m}{\lambda} - \frac{3}{8} \right] \frac{q^2}{m^2} + \frac{\alpha}{4m\pi} \sigma_{\mu\nu} q_\nu. \tag{4.89}$$

The first term in (4.89) was calculated by Feynman [4], the second by Schwinger [3]. The latter term is just the anomalous magnetic moment $\alpha/2\pi$ of the electron.

Now we can calculate the experimentally measured cross section for the massive photon. The result turns out to be independent of the auxiliary photon mass λ if this mass is smaller than the threshold of sensitivity of our detector, $\lambda \ll \omega_{\min}$:

$$d\sigma = d\sigma_s + d\sigma_\gamma(\omega < \omega_{\min}) \quad \propto \quad \ln\frac{m}{\lambda} + \ln\frac{\lambda}{\omega_{\min}} = \ln\frac{m}{\omega_{\min}}.$$

The physical answer depends on the experimental energy resolution, and this solves the problem of infrared divergence.

4.8 The Dirac equation in an external field

We have calculated the first order corrections to scattering of electrons by an external field:

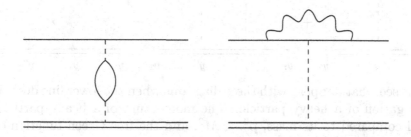

In certain situations it is necessary to include many higher order corrections. For example, for electron scattering off a nucleus with a large atomic number Z the rôle of the expansion parameter is played not by α but by $Z\alpha$, and $Z\alpha$ may be large for a heavy nucleus. Fortunately, the situation is somewhat simplified by the fact that all particles with large Z have large masses $M \gg m_{\mathrm{e}}$, and in the leading approximation we can neglect recoil corrections of order m/M.

We consider the interaction of a heavy charged particle with an electron and try to find all corrections in $Z\alpha$, neglecting recoil. Interaction of a light and a heavy particle (shown by the double line) is described by the

diagrams

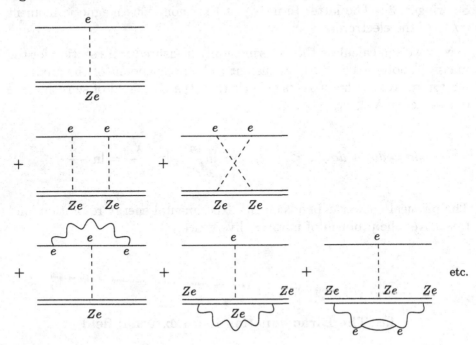

We start with the three pure exchange graphs:

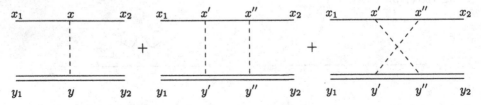

Let us see what happens with these diagrams when the lower line describes propagation of a heavy particle. The momentum of a heavy particle is small compared to its mass, $\mathbf{p}^2 \ll M^2$, and the free Green function of a heavy particle with spin $\frac{1}{2}$ (2.56),

$$G(y) = \int \frac{d^4p}{(2\pi)^4 i} e^{-ipy} \frac{M + \hat{p}}{M^2 - p^2 - i\delta}, \tag{4.90}$$

may be simplified.** For the heavy particle we write p_0 as

$$p_0 = M + \varepsilon,$$

** Unlike (2.56) we use $i\delta$ instead of $i\epsilon$ here to describe how the poles are shifted from the real axis, because ϵ is reserved as a standard notation for the binding energy in a bound state problem.

where $\varepsilon \ll M$ plays the rôle of the kinetic energy. Then

$$M^2 - p^2 = M^2 - (M + \varepsilon)^2 + \mathbf{p}^2 = -2M\varepsilon + \mathbf{p}^2 \simeq 2M\left(\frac{\mathbf{p}^2}{2M} - \varepsilon\right),$$

and

$$
\begin{aligned}
G(y) &= \int \frac{d\varepsilon d^3 p}{(2\pi)^4 i} \frac{e^{-i(M+\varepsilon)\tau}[M(1+\gamma_0) + \varepsilon\gamma_0 - \mathbf{p}\cdot\boldsymbol{\gamma}]e^{i\mathbf{p}\cdot\mathbf{y}}}{-2M\varepsilon - i\delta + \mathbf{p}^2} \\
&= \frac{e^{-iM\tau}}{2M}\theta(\tau)\int \frac{d^3p}{(2\pi)^3}e^{-i\frac{\mathbf{p}^2}{2M}\tau}e^{i\mathbf{p}\cdot\mathbf{y}}\left[M(1+\gamma_0) + \frac{\mathbf{p}^2}{2M}\gamma_0 - \mathbf{p}\cdot\boldsymbol{\gamma}\right] \quad (4.91) \\
&= \theta(\tau)e^{-iM\tau}\frac{1+\gamma_0}{2}\,\delta(\mathbf{y}).
\end{aligned}
$$

This is a natural result which simply means that up to corrections of the order of $\mathbf{p}^2/M^2$ the heavy particle propagates forward only in time $\tau \equiv y_0$ and stays practically at rest. Then in the diagram

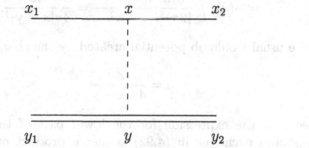

the heavy-particle line with an attached exchange photon is described by the expression

$$
\begin{aligned}
Ze \int &G(y_2 - y)i\gamma_\mu G(y - y_1)D(x - y)d^4y \\
&= iZe\delta(\mathbf{y}_2 - \mathbf{y}_1)\frac{1+\gamma_0}{2}\gamma_\mu\frac{1+\gamma_0}{2} \\
&\quad \times \int d\tau e^{-i(\tau_2-\tau)M-i(\tau-\tau_1)M}D(x-y)\theta(\tau_2 - \tau)\theta(\tau - \tau_1) \qquad (4.92) \\
&= \delta(\mathbf{y}_2 - \mathbf{y}_1)g_{\mu 0}e^{-iM(\tau_2-\tau_1)}[-u(\mathbf{x} - \mathbf{y}_1)]\frac{1+\gamma_0}{2},
\end{aligned}
$$

where

$$u = -iZe\int_{\tau_1}^{\tau_2} d\tau D(t - \tau, \mathbf{x} - \mathbf{y}_1)\theta(\tau_2 - \tau_1). \qquad (4.93)$$

In the derivation of (4.92) we used the trivial identities

$$\frac{1+\gamma_0}{2}\gamma_i\frac{1+\gamma_0}{2} = \frac{1+\gamma_0}{2}\frac{1-\gamma_0}{2}\gamma_i = 0,$$

$$\frac{1+\gamma_0}{2}\gamma_0\frac{1+\gamma_0}{2} = \frac{1+\gamma_0}{2},$$

so that

$$\frac{1+\gamma_0}{2}\gamma_\mu\frac{1+\gamma_0}{2} = g_{\mu 0}\frac{1+\gamma_0}{2}. \tag{4.94}$$

Let us calculate the integral (4.93) at $\tau_1 \to -\infty$ and $\tau_2 \to +\infty$. Using the explicit form of the photon propagator

$$D(t - \tau, \mathbf{x} - \mathbf{y}_1) = \int \frac{d^4k}{(2\pi)^4 i}\frac{e^{-ik_0(t-\tau)+i\mathbf{k}\cdot(\mathbf{x}-\mathbf{y}_1)}}{k_0^2 - \mathbf{k}^2},$$

we obtain

$$
\begin{aligned}
u &= -iZe\int_{-\infty}^{\infty} d\tau \int \frac{d^4k}{(2\pi)^4 i}\frac{e^{-ik_0(t-\tau)+i\mathbf{k}\cdot(\mathbf{x}-\mathbf{y}_1)}}{k_0^2 - \mathbf{k}^2} \\
&= -iZe\int \frac{d^4k}{(2\pi)^4 i}\frac{e^{-ik_0 t+i\mathbf{k}\cdot(\mathbf{x}-\mathbf{y}_1)}}{k_0^2 - \mathbf{k}^2}\, 2\pi\delta(k_0) \\
&= -iZe\int \frac{d^3k}{(2\pi)^3 i}\frac{e^{i\mathbf{k}\cdot(\mathbf{x}-\mathbf{y}_1)}}{-\mathbf{k}^2} = \frac{Ze}{4\pi|\mathbf{x}-\mathbf{y}_1|}.
\end{aligned}
$$

This is the usual Coulomb potential created by the charge Ze:

$$u = \frac{Ze}{4\pi|\mathbf{x}-\mathbf{y}|}. \tag{4.95}$$

We see that the expression for the lower part of the diagram with the one-photon exchange in (4.92) is just a product of the free heavy particle Green function (4.91) and the Coulomb potential (4.95). Only the Coulomb potential u in (4.92) contains the coordinate of the electron. Hence, the graph with one-photon exchange may be represented as a disconnected diagram, corresponding to two independent processes

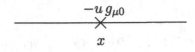

The double line describes free propagation of the heavy particle (4.91) while the upper line, with an attached cross marking the Coulomb potential (in the limit $\tau_1 \to -\infty$, $\tau_2 \to \infty$) describes the scattering amplitude of an electron in an external field created by the heavy particle with charge Ze:

$$\int d^4x\, G(x_2 - x)\left[-i\gamma_0 e\, u(x, y)\right] G(x - x_1).$$

Let us now turn to the ladder two-photon exchange:

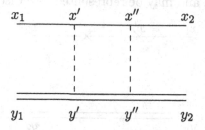

The double line with the attached photons now corresponds to the expression

$$\int G(y_2-y'')\,i\gamma_\mu Ze\,G(y''-y')\,i\gamma_\nu Ze\,G(y'-y_1)D(x'-y')D(x''-y'')d^4y'd^4y''$$

$$= ig_{\mu0}\,Ze\,\frac{1+\gamma_0}{2}\,ig_{\nu0}\,Ze\,\delta(\mathbf{y}_2-\mathbf{y}_1)\,e^{-iM(\tau_2-\tau_1)}$$

$$\times \int d\tau'd\tau''D(t'-\tau',\mathbf{x}'-\mathbf{y}_1)\,D(t''-\tau'',\mathbf{x}''-\mathbf{y}_1),$$

where the times ($\tau \equiv y_0$) on the heavy-particle line are ordered according to $\tau_1 < \tau' < \tau'' < \tau_2$. The time integrals do not factorize only because the integration variables are ordered by these inequalities. Happily the crossed ladder diagram

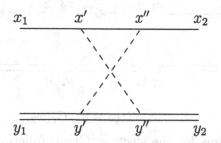

contains exactly the same analytic expression, but with the integration variables subject to the complementary restrictions $\tau_1 < \tau'' < \tau' < \tau_2$. Then the sum of these two diagrams produces an integral with no restrictions on the interaction times τ', τ'', and the integrals over τ', τ'' factorize. Each of these factorized integrals gives the Coulomb potential, and the sum of the ladder and crossed ladder diagrams is proportional to

$$\propto G(y_2-y_1)\,[-u(\mathbf{x}'-\mathbf{y}_1)]\,[-u(\mathbf{x}''-\mathbf{y}_1)].$$

Hence, the scattering amplitude with two-photon exchange is again a product of two factors, and may be represented as a disconnected diagram

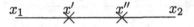

The lower line describes, as before, free motion of the heavy particle, and the upper part describes an independent process of double scattering of the electron by the Coulomb source.

Factorization into a free heavy particle propagation and an electron scattering in the external field is replicated in the sum of the diagrams with any fixed number of photon exchanges. This factorization simply means that the heavy particle neither experiences recoil, nor produces retardation.

Thus the total amplitude for electron scattering by a heavy particle may be represented as a sum of the factorized contributions, which reduces to the product of the electron Green function in an external field G_{e} and the free heavy particle propagator:

$$= G(y_2 - y_1)\, G_{\mathrm{e}}(x_2, x_1; y_1). \qquad (4.96)$$

Summing the diagrams for the electron Green function in the external field we immediately come to a graphical equation

$$(4.97)$$

where the bold line represents the electron Green function in the external Coulomb field. Analytically equation (4.97) reads

$$G_e(x_2, x_1; \mathbf{y}_1) = G(x_2 - x_1)$$
$$+ \int d^4x \, G(x_2 - x)[-ie\gamma_0 u(\mathbf{x}, \mathbf{y}_1)]G_e(x, x_1; \mathbf{y}_1). \qquad (4.98)$$

This integral equation may be easily converted into a differential equation with the help of the Dirac equation (2.54) for the free electron Green function:

$$\left(i\gamma_\mu \frac{\partial}{\partial x_\mu} - m\right) G(x) = i\delta(x).$$

Acting on (4.98) with the operator $i\gamma_\mu \frac{\partial}{\partial x_{2\mu}} - m$, we get

$$\left(i\gamma_\mu \frac{\partial}{\partial x_{2\mu}} - m\right) G_e(x_2, x_1; \mathbf{y}_1) = i\delta(x_2 - x_1)$$

$$+ e\gamma_0 u(\mathbf{x}_2, \mathbf{y}_1)\, G_e(x_2, x_1; \mathbf{y}_1),$$

or

$$\left(i\gamma_\mu \frac{\partial}{\partial x_{2\mu}} - m - e\gamma_0 u\right) G_e(x_2, x_1; \mathbf{y}_1) = i\delta(x_2 - x_1). \qquad (4.99)$$

The Green function may be represented as (compare (2.58), (2.59))

$$G_e(x_2, x_1; \mathbf{y}_1) = \begin{cases} \displaystyle\sum_n \Psi_n^+(x_2)\Psi_n^{+*}(x_1), & t_2 > t_1; \\ \displaystyle-\sum_n \Psi_n^-(x_2)\Psi_n^{-*}(x_1), & t_2 < t_1, \end{cases}$$

where $\{\Psi_n\}$ is a complete set of normalized solutions of the Dirac equation in the Coulomb field

$$\left\{i\gamma_\mu \frac{\partial}{\partial x_\mu} - m - \gamma_0 \frac{Ze^2}{4\pi|\mathbf{x} - \mathbf{y}_1|}\right\} \Psi_n(x) = 0. \qquad (4.100)$$

So far, we have collected all corrections in $Z\alpha$ generated by the diagrams with multiphoton exchanges. These corrections are effectively described by the diagrams

etc.

which can be summed with the help of the electron Green function in the Coulomb field.

There exist, however, other large contributions, for example self-energy corrections to the heavy line. In the first order in Z^2e^2 we have

$$= G(y_2 - y_1)\left(-Z^2e^2 \int_{\tau_1}^{\tau_2} d\tau' d\tau'' D_{00}(\tau'' - \tau')\right) \quad (4.101)$$

$$\tau_1 < \tau' < \tau'' < \tau_2;$$

in the next order, $(Z^2e^2)^2$,

etc.

As in the case of multiphoton exchanges between the heavy particle and the electron, summation of diagrams with a given number of photons attached in all possible orders effectively lifts restrictions on relative times and, as a result, integrations over the time coordinates of each self-energy insertion go independently of one another. Then the integrals again factorize and turn into products of a free Green function of the heavy particle and the integrals describing self-energy corrections. This happens for any number of self-energy insertions.

For example, for two one-loop self-energy insertions we obtain

$$G(y_2 - y_1)\left(-Z^2e^2 \int_{\tau_1}^{\tau_2} d\tau' d\tau'' D_{00}(\tau'' - \tau')\right)^2 \frac{1}{2!}, \quad \tau' < \tau''. \quad (4.102)$$

The factor $1/2!$ arises here because the number of topologically different diagrams is 2! times less than the number of photon permutations. For n one-loop self-energy insertions we similarly obtain

$$G(y_2 - y_1)\left(-Z^2e^2 \int_{\tau_1}^{\tau_2} d\tau' d\tau'' D_{00}(\tau'' - \tau')\right)^n \frac{1}{n!}, \quad \tau' < \tau''. \quad (4.103)$$

Then we can sum all these contributions:

$$G(y_2 - y_1)e^{-Z^2e^2 \int_{\tau_1}^{\tau_2} d\tau' d\tau'' D_{00}(\tau'' - \tau')}, \quad \tau' < \tau''. \quad (4.104)$$

Changing the integration variables $\tau = \tau'' - \tau'$, $x = \tau'' + \tau'$, and keeping only the leading contribution $\propto (\tau_2 - \tau_1)$ we obtain

$$G(y_2 - y_1)e^{-(\tau_2 - \tau_1)Z^2e^2 \int_0^{\tau_2 - \tau_1} d\tau D_{00}(\tau)}. \quad (4.105)$$

It is easy to see that in the limit $\tau_2 - \tau_1 \to \infty$ the integral in the exponent is purely imaginary:

$$\int_0^\infty d\tau \int \frac{d^4k}{(2\pi)^4 i} \frac{e^{-ik_0\tau}}{k^2 - i\varepsilon} = \int_0^\infty d\tau \int \frac{d^3k\, e^{-i|k|\tau}}{2|k|(2\pi)^3}.$$

After time integration

$$\int_0^\infty d\tau e^{-i|k|\tau} = \frac{1}{i|k|}$$

we obtain

$$Z^2 e^2 \int_0^\infty d\tau\, D_{00}(\tau) = -i Z^2 e^2 \int \frac{d^3k}{2|k|^2} \frac{1}{(2\pi)^3} \equiv i\delta M. \tag{4.106}$$

Hence, the exponent in the Green function of the heavy particle shifts after the inclusion of the self-energy corrections:

$$G(y_2 - y_1) \propto e^{-i(M+\delta M)(\tau_2-\tau_1)}. \tag{4.107}$$

Physically this means that the self-energy corrections renormalize the heavy particle mass. (In our non-relativistic approximation the δM is formally linearly divergent.)

Self-energy corrections to the heavy particle Green function may also contain a closed electron–positron loop

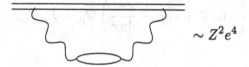

$$\sim Z^2 e^4$$

and the particles in the closed loops can themselves interact with the heavy particle

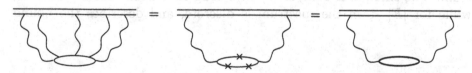

Summation of any number of these interactions results in the substitution of the electron Green functions in the external field for the free electron Green functions in the electron–positron loop. Then the heavy particle self-energy correction will include the photon Green function with this dressed electron–positron pair in the Coulomb field,

$$D_{00}^c(\tau) = \quad \text{-------------} \quad + \quad \text{-------} \bigcirc \text{-------}$$

instead of the free photon Green function

$$D_{00}(\tau) = \quad \text{-----------}$$

This again changes only the heavy particle self-energy and does not affect propagation of the scattering electron.

We conclude that even after including higher order corrections to heavy particle propagation all our integrals still factorize, and electron scattering off a heavy particle reduces effectively to two separate problems: electron scattering in the external field, and heavy particle mass renormalization.

4.8.1 Electron in the field of a supercharged nucleus

The energy spectrum of an electron in an external field is determined by the Dirac equation (4.100). To find the spectrum we look for the energy eigenstates in the form $\Psi_n = \exp(-iE_n t)\Psi_n(\mathbf{r})$, substitute them in (4.100), and multiply by γ_0. Then we arrive at the stationary Dirac equation for a particle in an external field[††]

$$\left(-i\boldsymbol{\alpha}\boldsymbol{\nabla} + m\gamma_0 - \frac{Ze^2}{4\pi r}\right)\Psi_n = E_n\Psi_n, \qquad (4.108)$$

where $\boldsymbol{\alpha} = \gamma_0\boldsymbol{\gamma}$.

The Dirac equation has both discrete and continuous parts of the spectrum. The energies $E > m$, $E < -m$ belong to the continuous spectrum, while for $|E| < m$ the spectrum is discrete, and describes the bound

---

[††] Note the change of sign of the potential in comparison with (4.100). In the derivation of (4.100) we have ascribed the charges e and Ze to the light and heavy particle, respectively. This means that we implicitly assumed that both the electron and the heavy particle have charge of the same sign. In real life the electron is charged negatively and the nucleus positively. This means that the Coulomb potential enters (4.100) with the opposite sign. Below we will always write the Coulomb potential in the Dirac equation with the sign corresponding to attraction, that is, opposite to that in (4.100).

states.

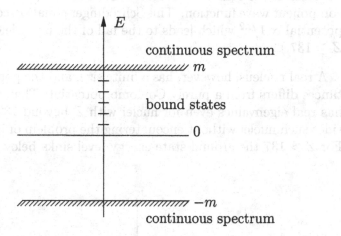

According to the Dirac equation, the energy E of the lowest bound state in the Coulomb field is

$$\frac{E}{m} = \sqrt{1 - (\alpha Z)^2}. \tag{4.109}$$

For the binding energy ε $(E = m + \varepsilon)$ we then have

$$\frac{\varepsilon}{m} = \sqrt{1 - (\alpha Z)^2} - 1.$$

The energy of the ground state decreases with Z

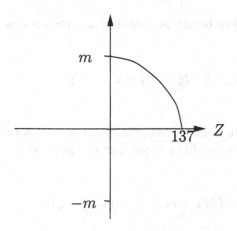

and becomes zero at $Z = 137$. The Dirac equation with the Coulomb potential has no sensible ground state solution for $Z \geq 137$, since the energy becomes imaginary. This is a special property of the purely Coulomb potential. (It can be understood if we represent the Dirac equation in the form of the equivalent second-order Schrödinger equation for the two-component wave function. This Schrödinger equation contains an effective potential $\propto 1/r^2$ which leads to the fall of the particle onto the centre for $Z \geq 137$.)

A real nucleus, however, has a finite size, and the potential at small distances differs from a purely Coulomb potential. Then the Dirac equation has real eigenvalues even for nuclei with Z beyond 137, and we can consider such nuclei without encountering the problem of imaginary energies. For $Z > 137$ the ground state energy level sinks below zero

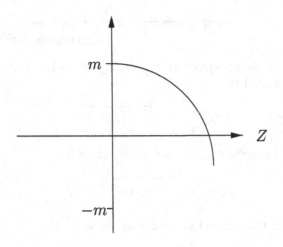

This means that the atom becomes lighter than the nucleus

$$M_A = M_Z + m_{\mathrm{e}} + \varepsilon < M_Z.$$

As long as $E > -m$, the nucleus remains stable since its mass is smaller than the sum of the masses of the atom and the positron

$$M_Z - (M_A + m_{\mathrm{e}}) = -2m_{\mathrm{e}} - \varepsilon < 0.$$

Increasing Z further, we reach a value Z_{cr} for which the binding energy becomes equal to $-2m_{\mathrm{e}}$ ($E = -m_{\mathrm{e}}$). At $Z = Z_{\mathrm{cr}}$ the decay of a nucleus into an 'atom' and a positron becomes energetically allowed (see Fig. 4.8)

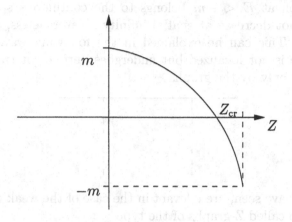

Fig. 4.8

and the process

$$Z \to (Ze^-) + e^+$$

should take place. Hence, for Z greater than critical, $Z > Z_{cr}$, the nucleus is an unstable system while the 'atom' (ion with a sub-critical charge $Z-1$) is stable.

One can easily understand the decay of a supercharged nucleus in terms of Green functions. The Green function for a heavy particle is proportional to the exponential of the self-energy correction (4.107)

$$G_Z(y_2 - y_1) \sim e^{-(\tau_2-\tau_1)\left(Z^2e^2 \int_0^\infty D_{00}(\tau)d\tau\right)}.$$

As we have seen, the free photon Green function generates a purely imaginary integral in the exponent. However, taking account of the radiatively corrected photon propagator which includes a contribution of the electron–positron pair in the external Coulomb field

the integral in the exponent at $Z > Z_{cr}$ acquires a real part, $i(\delta M + i\gamma)$. This additional real part corresponds to the creation of a real electron–positron pair in the field of the nucleus. As a result, the heavy particle Green function decays with time

$$G_Z \propto e^{-\gamma(\tau_2-\tau_1)}.$$

This damping reflects the decay of the heavy charged particle into an 'atom' and a positron.

The atom in the case of the supercharged nucleus remains stable, but becomes essentially a multiparticle system. In fact, any solution of the

Dirac equation at $E < -m$ belongs to the continuous spectrum and, hence, does not decrease at spatial infinity. Nevertheless, the localized atom exists. This can be explained in the following way. The single electron state is not localized but undergoes permanent transmutations described not only by the graphs

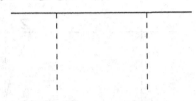

which, as we have seen, are relevant in the case of the weak coupling, but also by the so-called Z-graphs of the type

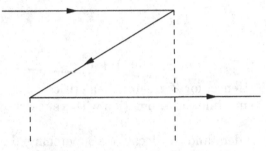

The electron in the supercharged atom does not preserve its indentity, it is delocalized and is continuously replaced, but these processes go in such a way that the charge remains localized. Clearly, such a system cannot be described in a single-particle framework.

Realistic estimates of the critical charge Z_{cr} give

$$Z_{cr} \simeq 170.$$

In principle, such charges could be created experimentally in collisions of two heavy ions when two nuclei come very close to each other. (For a more detailed description of the behaviour of electrons in critical fields see, e.g., Popov and Zeldovich [5], Migdal [6]).

4.9 Radiative corrections to the energy levels of hydrogen-like atoms. The Lamb shift

The case of high nuclear charge Z considered in the previous section is not the only situation when it is not sufficient to consider contributions of only a few lowest order diagrams. One should not forget that the diagrams are functions of the kinematical variables (energy, momentum transfer), and in certain kinematical regions these functions may become large, and compensate suppression provided by high powers of α.

Consider again electron scattering off a heavy particle

$$
\begin{array}{c}
\underline{\hspace{2cm}}\\[-2pt]
p_1 \qquad\qquad p_2 \\[4pt]
+
\end{array}
\quad
\begin{array}{c}
\overset{\textstyle Ze^2}{\underline{\hspace{2cm}}}\\[-2pt]
p_1 \quad\ \vdots\ q\quad\ p_2 \\[4pt]
+
\end{array}
\quad
\begin{array}{c}
\overset{\textstyle Ze^2\quad Ze^2}{\underline{\hspace{2.5cm}}}\\[-2pt]
p_1 \quad \vdots\ q_1\ \vdots\ q-q_1 \quad p_2
\end{array}
$$

The second diagram is just the leading order contribution to the Coulomb scattering, and contains the factor Ze^2/q^2 with q the momentum transfer. The third diagram is described by the integral

$$
(Ze^2)^2 \int \frac{d^3 q_1}{q_1^2} \frac{m + \hat{p}_1 + \hat{q}_1}{m^2 - (p_1 + q_1)^2} \frac{1}{(q - q_1)^2}.
$$

Let us estimate this two-photon integral for a non-relativistic incoming electron. In this case

$$
p_{10} \simeq m + E, \qquad E \equiv \frac{\mathbf{p}_1^2}{2m},
$$

and

$$
m^2 - (p_1 + q_1)^2 = m^2 - (m + E)^2 + (\mathbf{p}_1 + \mathbf{q}_1)^2 \simeq 2mE + (\mathbf{p}_1 + \mathbf{q}_1)^2.
$$

The integration volume for the non-relativistic integration momenta $q_1 \sim p_1$ is of the order of $\int d^3 q_1 \sim p_1^3$, and the integral can be estimated as

$$
\sim (Ze^2)^2 \frac{m}{p_1^3}.
$$

The sum of the one- and two-photon exchange diagrams is then, symbolically,

$$
\sim \frac{Ze^2}{p_1^2}\left(1 + \frac{Ze^2 m}{p_1}\right).
$$

The second term in the brackets, Ze^2/v, may become large if electron velocity $v = |\mathbf{p}_1|/m$ is sufficiently small.

Hence, dealing with non-relativistic electrons with momenta of the order of $p_1 \sim mZe^2$, we have to consider diagrams with any number of photon exchanges on the same footing and sum them exactly (even for $Z \sim 1$). As we already know, all orders in Ze^2 are embodied into the electron Green function in the external field, which satisfies the graphical equation (4.97)

$$
\underline{\hspace{3cm}} = \underline{\hspace{2.5cm}} + \underline{\hspace{1.2cm}}\times\underline{\hspace{1.2cm}}
$$

equivalent to the Dirac equation. By construction, solutions of this equation are exact in the small parameter Ze^2.

Consider now even smaller corrections of the order $(Ze^2)^n e^2$ to electron propagation. These corrections are generated by the processes with emission and absorption of additional photons by the electron, and by the processes with additional closed electron loops, like

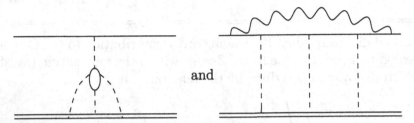

and

Let us see how these diagrams change the equation (4.97) for the Green function in external field.

First, the external potential is modified due to vacuum polarization

$$U(q) \longrightarrow U(q)\left[1 + \Pi^c(q^2)\right], \qquad U(q) = \frac{Ze^2}{q^2}, \qquad (4.110)$$

where $\Pi^c(q^2)$ is the photon polarization operator in the external field (i.e. constructed from the electron Green functions in the external field).

Second, a new term, electron self-energy in the external field (i.e. with the virtual electron line described by the external field electron Green function) arises in the equation due to photon emission and absorption by the electron. Hence, equation (4.97) for the electron Green function acquires the form

$$\underline{\qquad\qquad} = \underline{\qquad\qquad} + \underline{\qquad\qquad\overset{U(q)}{\times}\qquad\qquad} +$$

$$+ \underline{\qquad\overset{\frown}{\qquad}\qquad} + \underline{\qquad\overset{\Pi^c(q^2)U(q)}{\underset{\circ}{\qquad}}\qquad}$$

Let us see how the new terms in this equation affect the energy levels of the electron in light atoms with small Z, in particular in the hydrogen atom with $Z = 1$. The binding energy in light atoms

$$p_0 - m \sim \frac{\mathbf{q}^2}{2m} \sim \frac{Z^2\alpha^2 m}{2} \ll m$$

is small, and the bound electron is non-relativistic.

For the non-relativistic electron the self-energy correction in the external field simplifies. If spanning photons are sufficiently hard,[‡‡] we can treat the electron in the intermediate state as essentially free, since the energy $p_0 - k_0 - m$ of this virtual electron is large compared to the binding energy. Hence, for light atoms we can expand the electron self-energy in the external field in the number of exchanged photons

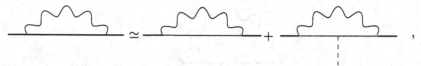

and ignore electron binding in the intermediate states. We see that for a sufficiently hard spanning photon the self-energy correction in the external field reduces to the one-loop self-energy and one-loop vertex of the free electron which we have already calculated.

Similarly, for light atoms we can ignore electron binding in the polarization loop as well

$$\simeq \Pi^c(q^c)U(q) \, ,$$

since the main contribution to the polarization integral comes from virtual electrons with high energies.

Let us estimate the relative magnitude of the polarization correction in comparison with the Coulomb potential U. According to (4.76)

$$\frac{\Pi^c U}{U} = \Pi^c \simeq -\frac{\alpha}{15\pi}\frac{q^2}{m^2} \sim \alpha(Z\alpha)^2, \qquad (4.111)$$

because $|q^2| \sim |\mathbf{p}|^2$ and for the bound electron $|\mathbf{p}| \sim mZ\alpha$.

We now turn to the correction generated by the vertex function

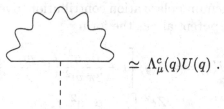

$$\simeq \Lambda_\mu^c(q)U(q) \, .$$

As we have seen before in (4.89), the vertex part near the mass shell $(p_1^2 \approx p_2^2 \approx m^2)$ has the form

$$\Lambda_\mu^c = \left[\frac{\alpha}{3\pi}\frac{q^2}{m^2} \left(\ln\frac{m}{\lambda} - \frac{3}{8} \right) + \alpha\frac{\hat{p}_1 - m}{m}C_1 + \alpha\frac{\hat{p}_2 - m}{m}C_2 \right] \gamma_\mu, \qquad (4.112)$$

[‡‡] The contribution of soft spanning photons will be considered separately below.

where we have omitted the anomalous magnetic moment contribution and added two contributions which arise because the bound atomic electron is slightly off mass shell. These are the first terms in the expansion of Λ_μ in powers of $(\hat{p}_1 - m)$ and $(\hat{p}_2 - m)$. In a light atom these corrections can be estimated as $\alpha(p_0 - m)/m \sim \alpha q^2/2m^2 \sim \alpha(Z\alpha)^2$.

Consider, finally, the correction induced by the self-energy operator

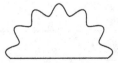

Near the mass shell $\hat{p} \simeq m$ we have (see (4.18))

$$\Sigma_c \sim \alpha \frac{(\hat{p} - m)^2}{m}.$$

To find the relative magnitude of the self-energy correction we compare it, as before, with the Coulomb potential contribution and derive the estimate

$$\frac{\Sigma_c}{\langle U \rangle} \sim \frac{\Sigma_c}{\hat{p} - m} \sim \frac{\alpha(\hat{p} - m)^2/m}{\hat{p} - m} = \alpha \frac{\hat{p} - m}{m} \sim \alpha \frac{q^2}{m^2} \sim \alpha(Z\alpha)^2.$$

This contribution is of the same order as all other corrections (4.111) and (4.112). It cancels, however, with the last two terms of Λ_μ^c in (4.7) due to the Ward identity

$$\Gamma_\mu(p, p) = -\frac{\partial G^{-1}}{\partial p}.$$

As a result, total correction to the energy levels of relative order $\alpha(Z\alpha)^2$ is generated by the sum of the contributions of the first term in Λ_μ^c (4.112) and of the vacuum polarization contribution given in (4.111). The respective effective potential has the form

$$U(q) \Longrightarrow U(q) \left[1 + \frac{\alpha}{3\pi} \frac{q^2}{m^2} \left(\ln \frac{m}{\lambda} - \frac{3}{8} - \frac{1}{5} \right) \right]$$

$$= -\frac{Ze^2}{q^2} \left[1 - \frac{\alpha}{3\pi} \frac{q^2}{m^2} \left(\ln \frac{m}{\lambda} - \frac{3}{8} - \frac{1}{5} \right) \right]. \tag{4.113}$$

The correction induced by the radiative effects is momentum independent, and corresponds therefore to a contact δ-functional potential in coordinate representation:

$$\tilde{u}(\mathbf{r}) = -\delta(\mathbf{r}) \frac{4}{3} \frac{Z\alpha^2}{m^2} \left[\ln \frac{m}{\lambda} - \frac{3}{8} - \frac{1}{5} \right]. \tag{4.114}$$

So the radiatively corrected equation for the electron Green function in the external field is

$$\left[i\gamma_\mu \frac{\partial}{\partial x_\mu} - m + \gamma_0 \left(\frac{Ze^2}{4\pi r} + \tilde{u}(r) \right) \right] G_{\mathrm{e}} = i\delta(x),$$

and the equation for the stationary wave function has the form

$$E\Psi = \left\{ \boldsymbol{\alpha} \cdot \mathbf{p} + m\gamma_0 - \frac{Z\alpha}{r} + \delta(\mathbf{r}) \frac{4}{3} \frac{Z\alpha^2}{m^2} \left[\ln \frac{m}{\lambda} - \frac{3}{8} - \frac{1}{5} \right] \right\} \Psi. \qquad (4.115)$$

Correction to the energy levels is given, as usual, simply by the matrix element of the perturbation potential between the unperturbed wave functions. To calculate this correction we will use non-relativistic Schrödinger wave functions of the hydrogen-like atom, since relativistic corrections to these functions contain higher powers of $(Z\alpha)^2$ and are small for $Z \sim 1$. The matrix element of the δ-like potential is proportional to $|\Psi(0)|^2$, and we easily obtain

$$\Delta E_{nj\ell} = \frac{4}{3} \frac{Z\alpha^2}{m^2} \left[\ln \frac{m}{\lambda} - \frac{3}{8} - \frac{1}{5} \right] |\Psi_{njl}(0)|^2, \qquad (4.116)$$

where n is the principal quantum number, j is the total angular momentum, and ℓ the orbital angular momentum of the electron. In the non-relativistic approximation only the wave functions of the S-states (i.e. states with $\ell = 0$) do not vanish in the origin. At $r = 0$ the value of the Schrödinger–Coulomb wave function squared is

$$|\Psi_{ns}(0)|^2 = \frac{1}{\pi} \left(\frac{\alpha Z m}{n} \right)^3, \qquad (4.117)$$

and

$$\Delta E_{ns} = \frac{4}{3} \frac{\alpha (Z\alpha)^4 m}{n^3 \pi} \left[\ln \frac{m}{\lambda} - \frac{3}{8} - \frac{1}{5} \right], \qquad (4.118)$$

or, in terms of the Bohr ground state energy $E_{\mathrm{B}} = \alpha^2 m/2$,

$$\Delta E_{ns} = \frac{8}{3\pi} \frac{Z^4 \alpha^3}{n^3} \left[\ln \frac{m}{\lambda} - \frac{3}{8} - \frac{1}{5} \right] E_{\mathrm{B}}. \qquad (4.119)$$

Due to this correction the $2S_{1/2}$ energy level in one-electron atoms is shifted upwards with respect to the $2P_{1/2}$ level. This 'Lamb shift' was discovered experimentally by Lamb and Retherford [7] in 1947.

The expression (4.119) for the Lamb shift depends on the unphysical photon mass λ. It emerged in our description of electron scattering when we wanted to get rid of the infrared divergence at small photon frequencies. We have seen that creation of a charged particle is accompanied by production of a large number of soft photons, and the auxiliary photon mass disappears from the final result if one properly takes into account this accompanying radiation. Now, however, we consider an atom which is a neutral system and, hence, there is no real photon emission. This means that there is no reason why any photon mass should enter the expression for the radiative shift of energy levels.

What did we do wrong? Recall that the mass λ arose because we replaced the Green function of the bound electron by the free Green function. For very soft spanning photons this is obviously wrong. Such an approximation is valid for $k_0 \gg Z\alpha m$, but if k_0 is less than the binding energy, electron binding becomes essential. Fortunately, while emitting (and eventually absorbing) very soft photons with $k_0 \ll m$ the electron remains non-relativistic, and the respective contribution may be calculated in the framework of non-relativistic quantum mechanics.

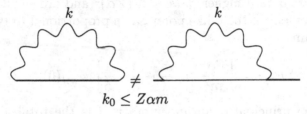

For light atoms $\alpha Z \ll 1$, and the regions $k_0 \gg Z\alpha m$ and $k_0 \ll m$ overlap. Then we can match the contributions obtained in two different ways and express λ in terms of parameters of the non-relativistic theory:

$$\ln \frac{m}{\lambda} = \ln \frac{m}{2\varepsilon_0} + \frac{5}{6},$$

where ε_0 is the average ionization potential for the atom (characterising typical binding energy).

The smallness of the parameter $Z\alpha$ is crucial for the validity of our calculations. In heavy atoms ($Z\alpha \sim 1$) the electron is relativistic, and the regions where different approximations work do not overlap. In such a case only numerical calculation of the Lamb shift is possible.

To obtain a complete expression for the Lamb shift in light hydrogen-like atoms, in the final result one also has to restore the anomalous magnetic moment contribution from (4.89). Then an additional term 3/8 arises in the brackets in (4.119), and we finally obtain

$$\Delta E_{ns} = \frac{4}{3\pi} \frac{\alpha (Z\alpha)^4 m}{n^3} \left[\ln \frac{m}{2\varepsilon_0} + \frac{19}{30} \right]. \tag{4.120}$$

5

Difficulties of quantum electrodynamics

5.1 Renormalization and divergences

We have considered the exact electron Green function G (4.13),

$$G = \frac{1}{m_0 - \hat{p} + \Sigma(\hat{p})},$$

the exact photon Green function $D_{\mu\nu}$ (4.36),

$$D = \frac{1}{k^2[1 - \Pi(k^2)]},$$

and the exact vertex part Γ_μ (4.39), and demonstrated that all observables may be calculated in terms of these three functions.

As we discussed, the electron Green function G does not have a pole at $\hat{p} = m_0$, i.e. this 'bare' mass is only a formal parameter of the theory and is unobservable. Then we represented the electron Green function in the form (4.20),

$$G(p) = \frac{Z}{m - \hat{p} + \Sigma_c(\hat{p})} = Z_2 G^c(p),$$

where (see (4.19), (4.23))

$$\Sigma_c(\hat{p}) = \frac{\Sigma(\hat{p}) - \Sigma(m) - \Sigma'(m)(\hat{p} - m)}{1 - \Sigma'(m)} \approx (\hat{p} - m)^2,$$

$$Z_2 = \frac{1}{1 - \Sigma'(m)}.$$

We concluded from these equations that the electron Green function has a pole at $\hat{p} = m$, with m the physical (or *renormalized*) mass.

Similarly, we wrote the photon Green function as (4.38)

$$D(k^2) = \frac{Z_3}{k^2[1 - \Pi_c(k^2)]} = Z_3 D^c(k^2),$$

241

where (4.37)

$$\Pi_c(k^2) = \frac{\Pi(k^2) - \Pi(0)}{1 - \Pi(0)} \sim k^2,$$

$$Z_3 = \frac{1}{1 - \Pi(0)}.$$

This means that the renormalized photon mass remains equal to zero.

For the vertex part we obtained (4.43)

$$\Gamma_\mu = Z_1^{-1}\Gamma_\mu^c, \quad \Gamma_\mu^c(m, m) = \gamma_\mu.$$

Moreover, we have shown in Section 4.5 that all graphs may be written in terms of the renormalized functions Γ_μ^c, G^c and D^c, exactly in the same way as in terms of the bare (unrenormalized) functions Γ_μ, G and D, if one substitutes the renormalized charge (4.48)

$$e_c^2 = Z_3 e^2$$

for the bare charge e (in this expression for the renormalized charge we used the Ward identity, $Z_1 = Z_2$). Thus, the amplitudes may be written exclusively in terms of physically observable renormalized charge and mass.

Can we calculate physical charge e_c and mass m in terms of the bare ones? The answer is no, since the respective integrals turn out to be divergent. Since we were forced to introduce the renormalized functions G^c, D^c and Γ^c anyway, the whole scheme of quantum electrodynamics would make sense if we could prove that these functions depend neither on the bare mass, nor on the bare charge. To this end we need to find such equations for Γ^c, Σ^c, Π^c which include only renormalized mass and charge and not the bare ones. Then all observables would depend only on the physical charge and mass, and the *renormalization procedure* would make sense.

5.1.1 Divergences of Feynman diagrams

Let us first consider what divergences exist in quantum electrodynamics and what is their origin. It is easy to see that there are three types of divergences in our theory:

(1) ultraviolet divergences which arise when the integrals diverge at large integration momenta $k \to \infty$,

(2) infrared divergences which are due to singularities of the integrands at small integration momenta $k \to 0$,

(3) possible poles and other singularities of the amplitudes which depend on external momenta.

We discussed the physical meaning of infrared divergences in Sections 2.9 and 4.7, and came to the conclusion that they are absent if the problems in QED are properly formulated. As to divergences of the third type, it is clear that the amplitudes may be singular for certain values of the external momenta, for example, when momentum k of a certain propagator in the diagram turns out to be on the mass shell $k^2 = m^2$. We can prove, however, that these singularities are absent if all external momenta are space-like and satisfy the triangle inequality. This is because in such a situation the contours of integration in the Feynman integrals may be rotated as shown in Fig. 5.1, so that all integration momenta become Euclidean:

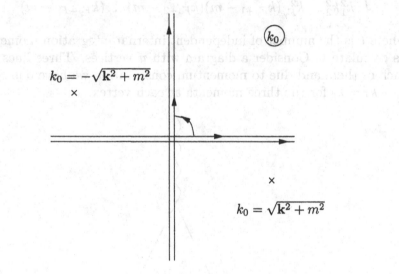

Fig. 5.1

The physical scattering amplitudes with time-like external momenta may be obtained from the amplitudes with Euclidean external momenta by analytic continuation. All singularities of the amplitudes as functions of the external momenta arise after this continuation. They are physically meaningful, connected with the unitarity condition (see discussion in Chapter 3), and we will not discuss them here.

Still, there remains the problem of the ultraviolet divergences, which survive even when the integration momenta are Euclidean.

Consider an arbitrary skeleton diagram

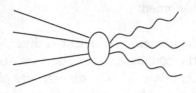

Due to current conservation it has an even number of external electron
lines. Let F_e and F_γ be the number of internal electron and photon
lines, and N_e and N_γ the number of external electron and photon lines,
respectively. The amplitude that corresponds to our diagram is given by
the integral

$$\int \frac{d^4 k_1 \ldots d^4 k_\ell}{k_1^2 k_2^2 \ldots k_{F_\gamma}^2 (\hat{k}_{F_\gamma+1} - m)(\hat{k}_{F_\gamma+2} - m) \ldots (\hat{k}_{F_\gamma+F_e} - m)}, \qquad (5.1)$$

where ℓ is the number of independent internal integration momenta. Let
us calculate ℓ. Consider a diagram with n vertices. Three lines meet at
each of them and due to momentum conservation we have one condition
$k_i + k_j = k_\ell$ for the three momenta at each vertex:

In fact, the number of conditions is $n - 1$ rather than n, since one of the
conditions corresponds to the conservation of the total four-momentum
and does not restrict the internal momenta. Hence, the number of independent internal momenta is

$$\ell = F_e + F_\gamma - n + 1.$$

The integral (5.1) diverges if the overall power of the differentials is larger
than or equal to the power of the denominator. This means that the
integral is convergent only if

$$4\ell - 2F_\gamma - F_e = 3F_e + 2F_\gamma - 4n + 4 < 0. \qquad (5.2)$$

We now show that this difference is independent of the number of internal lines (F_e and F_γ) and depends only on the number of external lines or, in other words, on the physical process itself. Consider first a closed internal electron line (see Fig. 5.2).

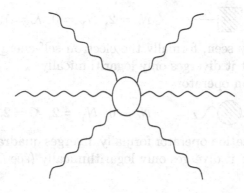

Fig. 5.2

For such a closed electron line the number of vertices n is equal to the number of segments of the electron line between the vertices

$$n = F_e.$$

An open electron line with n vertices on it contains $n-1$ intervals between the vertices plus two external lines. Hence, for an arbitrary diagram with n vertices we have

$$n = F_e + \frac{N_e}{2}.$$

Each internal photon line connects two vertices, and each external photon line ends at a vertex. Hence,

$$n = 2F_\gamma + N_\gamma.$$

Thus, the number of external lines and vertices determines the number of internal lines:

$$F_e = n - \frac{N_e}{2}, \quad F_\gamma = \frac{n}{2} - \frac{N_\gamma}{2}.$$

Then the condition for convergence of the integral for the amplitude becomes

$$K = 3F_e + 2F_\gamma - 4n + 4 = -\frac{3}{2}N_e - N_\gamma + 4 < 0, \qquad (5.3)$$

and convergence depends only on the number of external lines. We see that the graphs describing complicated processes with a large number of external lines are convergent. Hence, to learn everything about ultraviolet divergences it suffices to list and study the diagrams with a small number of external lines.

Let us consider the simplest graphs.

(1) Electron self-energy,

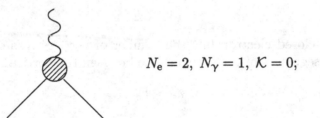

$N_e = 2,\ N_\gamma = 0,\ \mathcal{K} = 1.$

As we have already seen, formally the electron self-energy is linearly divergent, but in fact it diverges only logarithmically.

(2) The polarization operator

$N_e = 0,\ N_\gamma = 2,\ \mathcal{K} = 2.$

The photon polarization operator formally diverges quadratically, but due to gauge invariance it diverges only logarithmically (see Chapter 4).

(3) The vertex part

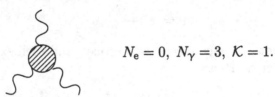

$N_e = 2,\ N_\gamma = 1,\ \mathcal{K} = 0;$

The vertex part diverges logarithmically.

(4) The three-photon diagram

$N_e = 0,\ N_\gamma = 3,\ \mathcal{K} = 1.$

Diagrams of this type are formally linearly divergent. However, we will see that such diagrams vanish due to charge conjugation invariance.

(5) Compton effect

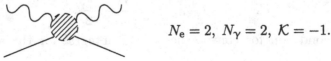

$N_e = 2,\ N_\gamma = 2,\ \mathcal{K} = -1.$

The diagram is convergent.

(6) Fermion–fermion scattering

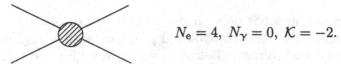

$N_e = 4,\ N_\gamma = 0,\ \mathcal{K} = -2.$

The diagram is convergent.
(7) Light by light scattering

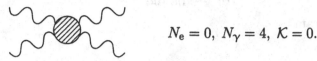

$$N_e = 0, \ N_\gamma = 4, \ \mathcal{K} = 0.$$

Formally the diagram looks logarithmically divergent, but due to gauge invariance it actually converges.

Let us prove the Furry theorem, which states that the three-photon vertex (as well as any other graph with an odd number of external photons only) is zero

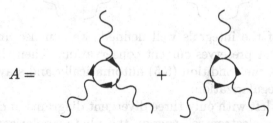

$$\equiv 0$$

We know that quantum electrodynamics is charge conjugation invariant. As we have seen, charge conjugation changes the sign of the photon wave function: $e_\mu \rightarrow -e_\mu$, therefore an amplitude with an odd number of photons also changes sign. The only difference between the Feynman diagrams with external photons before and after charge conjugation is that the directions of the internal fermion lines change. But how can the amplitude know whether there is a particle or an antiparticle inside? Such a substitution cannot change the amplitude and, hence, the amplitude has to vanish identically. Let us illustrate this by the example of the simplest process:

$$A = \qquad + \qquad$$

Under charge conjugation the first diagram changes sign and turns into the second diagram, which means that their sum is identically zero.

Now consider light by light scattering

$$= M_{\mu_1 \mu_2 \mu_3 \mu_4}(k_1, k_2, k_3, k_4). \tag{5.4}$$

Due to current conservation we have for any external photon momentum

$$k_1^{\mu_1} M_{\mu_1\mu_2\mu_3\mu_4} = 0. \qquad (5.5)$$

But the diagram

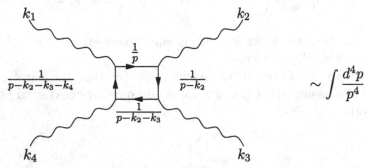

is logarithmically divergent at large integration momenta p. Since in the divergent part of the diagram the external momenta k_i in the denominators can be ignored, it turns out to be independent of k_i. As a result, the divergent part of the amplitude (5.4) does not satisfy the current conservation condition (5.5) because it is impossible to get zero after multiplying an amplitude, which does not depend on k, by k_μ. According to the Feynman rules, we have to sum three diagrams

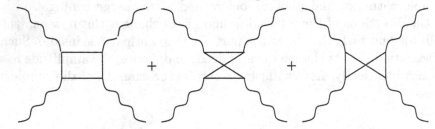

To make each of the integrals well defined, we can use any ultraviolet regularization that preserves current conservation. Then the sum of the integrals satisfies the condition (5.5) automatically and stays finite when we remove the regularization.

Hence, we are left with only three divergent diagrams in quantum electrodynamics: the electron self-energy, the photon polarization operator and the vertex part

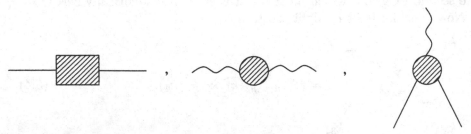

5.1.2 Renormalization

Let us see how renormalization removes all three divergences in electrodynamics. We assume temporarily that

$$G = G_0, \ D = D_0,$$

and try to construct an equation for the vertex part. By definition,

$$\text{(5.6)}$$

This is, in fact, an equation for Γ_μ. Indeed, (5.6) may be represented as a relationship for the skeleton diagrams

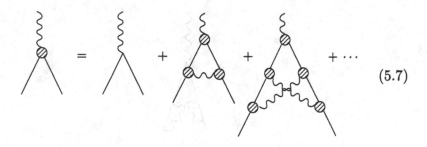

$$\text{(5.7)}$$

Iterating (5.7), we obtain (5.6). It is important to realise that (5.7) does not contain ladder diagrams of the type

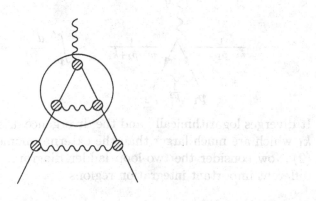

because the vertex part in the circle is already included in the graph

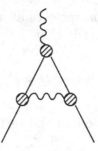

The ladder diagram

and the diagram with crossed photon lines

diverge differently.

(1) Consider first the one-loop diagram

$$\frac{1}{m-\hat{p}_1+\hat{k}_1} \qquad \frac{1}{m-\hat{p}_2+\hat{k}_1} \qquad \sim \int_p^\Lambda \frac{d^4 k_1}{k_1^4} \sim \ln \frac{\Lambda^2}{p^2}.$$

with labels q, p_1, $\frac{1}{k_1^2}$, p_2.

It diverges logarithmically and the divergence arises at internal momenta k_1 which are much larger than the external momenta $k_1 \gg p_1, p_2, m$.

(2) Now consider the two-loop ladder diagram. This diagram contains different important integration regions.

$$q$$

$$p_1 - k_1 - k_2 \qquad p_2 - k_1 - k_2$$

$$p_1 - k_1 \qquad k_2 \qquad p_2 - k_1$$

$$k_1$$

$$p_1 \qquad p_2$$

(a) If momentum k_2 is fixed, this diagram is convergent. For $k_1 \gg k_2$ we have

$$\int_{k_2}^{\infty} \frac{d^4 k_1}{k_1^8} \sim \frac{1}{k_2^2},$$

and integration over k_2 in this region produces a logarithm of the only momentum relevant for the process

$$\int \frac{d^4 k_2}{k_2^4} \sim \ln q.$$

(b) If, on the contrary, $k_2 \gg k_1$ (where k_1 is fixed), integration over k_2 gives a divergent integral

$$\int_{k_1}^{\Lambda} \frac{d^4 k_2}{k_2^4} \sim \ln \frac{\Lambda^2}{k_1^2}.$$

Then integration over k_1 leads to a double logarithm contribution

$$\int_{p}^{\Lambda} \frac{d^4 k_1}{k_1^4} \ln \frac{\Lambda^2}{k_1^2} \sim \ln^2 \frac{\Lambda^2}{p^2}.$$

The divergence is now more severe than in the one-loop case

and this is quite reasonable since in this integration region the two-loop ladder vertex contains a divergent one-loop subdiagram, and the last integration reduces to the same divergent integral, as in the one-loop case.

(3) Consider the two-loop diagram with crossed photon lines

$$q$$

$$p_1 - k_1 - k_2 \qquad p_2 - k_1 - k_2$$

$$p_1 - k_1 \qquad p_2 - k_2$$

$$k_1 \quad k_2$$

$$p_1 \qquad p_2$$

Now the integral converges both in the case where the integration goes over k_2 at a fixed value of k_1 ($k_2 \gg k_1$) and if it goes over k_1 with k_2 fixed ($k_1 \gg k_2$). The integral diverges only if $k_1 \sim k_2$, and this integration region leads to a logarithmic divergence

$$\int \frac{d^4 k}{k^4} \sim \ln \frac{\Lambda^2}{p^2}.$$

Repeating these considerations for more complicated diagrams, we can check that all terms in the skeleton expansion (5.7) diverge only logarithmically. As we have seen, the logarithmic divergences of the lowest order diagrams may be eliminated with the help of one subtraction. Hence, the subtraction procedure (renormalization) leads to a finite result, and it is possible to write such an equation for the vertex part which does not contain divergences at all.

Let us construct this finite equation for the vertex part. The total vertex part may be written as in (4.40)

$$\Gamma_\mu(p_1, p_2, q) = \gamma_\mu + \Lambda_\mu(p_1, p_2, q), \tag{5.8}$$

and according to (4.41) on the mass shell

$$\Lambda_\mu(m, m, 0) = \gamma_\mu \Lambda(m, m, 0). \tag{5.9}$$

Then the vertex part may be represented as

$$\Gamma_\mu(p_1, p_2, q) = \gamma_\mu(1 + \Lambda(m, m, 0)) + \Lambda_\mu(p_1, p_2, q) - \Lambda_\mu(m, m, 0). \tag{5.10}$$

In terms of the skeleton expansion, equation (5.10) has the form

$$\Gamma_\mu(p_1, p_2, q) = \gamma_\mu(1 + \Lambda(m, m, 0))$$

$$\tag{5.11}$$

The vertex part Γ_μ in (5.11) is present both on the left- and right-hand sides, so it is an integral equation for the vertex part (5.11). In terms of the renormalized vertex Γ_μ^c equation (5.11) has the form ($\Gamma_\mu^c = \Gamma_\mu Z_1$,

$Z_1^{-1} = 1 + \Lambda(m, m, 0)$, see (4.43), (4.42))

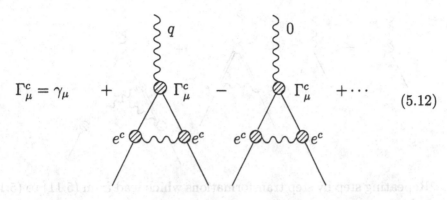

$$\Gamma_\mu^c = \gamma_\mu \quad + \quad \Gamma_\mu^c \quad - \quad \Gamma_\mu^c \quad + \cdots \tag{5.12}$$

where we used $e\Gamma_\mu = e_c\Gamma_\mu^c$. The renormalization constant Z_1^{-1} in the top vertices in all graphs in (5.11) is just a common factor which we cancelled in (5.12), while the renormalization constants in all other vertices are swallowed by the renormalized charge: $Z_1^{-1}e \to e_c$. As a result, all terms in (5.12) are ultraviolet finite because in each order of perturbation theory only the finite differences of divergent integrals of the type

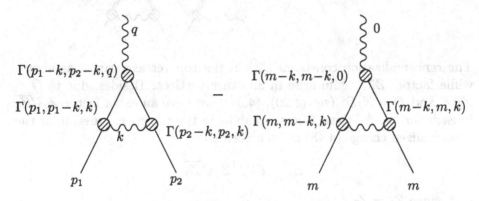

enter the right-hand side of this equation. Note that these differences go to zero when the integration momentum goes to infinity, $k \to \infty$, because

$$\Gamma(a, b, c) \to \Gamma(b, c) \qquad \text{at} \qquad b \sim c \gg a.$$

This we can prove with the help of the integral equation itself, or obtain directly from the perturbation theory graphs. This remarkable property of the vertices Γ means that at high external momenta they do not depend on the smallest momentum, and this guarantees that ultraviolet finiteness reproduces itself in higher order graphs.

We still have to obtain a finite equation for the vertex part in the real case with radiative corrections to the electron and photon Green functions taken into account. Instead of (5.11) we now have

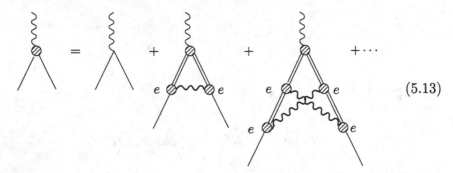

$$(5.13)$$

Repeating step by step transformations which lead from (5.11) to (5.12) we obtain

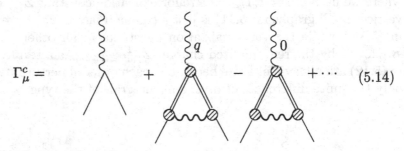

$$\Gamma_\mu^c = \qquad\qquad\qquad\qquad (5.14)$$

The renormalization constants Z_1^{-1} in the top vertices cancel as above, while factors Z_1^{-1} again arise in all other vertices. Besides, due to $G = Z_2 G^c$ and $D = Z_3 D^c$ (see (4.20), (4.38)) we have an extra factor $Z_2\sqrt{Z_3}$ in each vertex. All these factors combine in the correct expression for the renormalized charge (4.48) in each vertex

$$e_c = e\, Z_1^{-1} Z_2 \sqrt{Z_3}$$

and, since $Z_1 = Z_2$,

$$e_c^2 = Z_3 e^2.$$

Thus, we have obtained an integral equation for the vertex Γ_μ^c which contains only the renormalized charge e_c and the renormalized electron and photon Green functions G^c and D^c.

This integral equation would be finite if we had finite equations for the renormalized electron and photon Green functions. Let us derive such equations. It is easy to see that the electron Green function satisfies the equation

$$\underline{\quad G \quad} = \underline{\quad G_0 \quad} + \underline{\quad G_0 \quad} \boxed{-\Sigma(p)} \underline{\quad G \quad} \qquad (5.15)$$

where the electron self-energy is

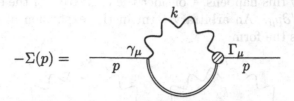

$$-\Sigma(p) =$$

Note that we have γ_μ instead of the exact vertex part in the left vertex, since all processes start with a simple photon emission

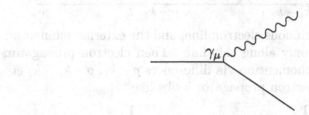

and only after this do all other processes take place. Analytically the *Schwinger–Dyson equation* for the electron self-energy has the form

$$-\Sigma(p) = e^2 \int \frac{d^4 k}{(2\pi)^4 i} \gamma_\mu \, G(p-k) \, \Gamma_\mu(p-k,p,k) \, D(k^2). \qquad (5.16)$$

A similar equation may be obtained for the photon polarization operator:

$$\Pi_{\mu\nu}(k^2) =$$

or, analytically,

$$\Pi_{\mu\nu}(k^2) = -e^2 \, \mathrm{Tr} \int \frac{d^4 p}{(2\pi)^4 i} \left\{ \gamma_\mu \, G(p) \, \Gamma_\nu(p+k,p,-k) \, G(p+k) \right\}. \qquad (5.17)$$

These equations, unfortunately, are not much help in the proof of ultraviolet finiteness of the renormalized vertex part since the integrals for $\Sigma(p)$ and $\Pi_{\mu\nu}(k^2)$ are ultravioletly divergent. From the point of view of convergence, we are interested, however, not in the self-energy $\Sigma(p)$ and the polarization operator $\Pi(k^2)$ themselves, but in differences of the type

$$\Sigma(p) - \Sigma(m) - \Sigma'(m)(\hat{p} - m)$$

which enter the expressions for the renormalized Green functions (see (4.19), (4.37)).

These two subtractions are just sufficient to eliminate all divergences. Let us see how this happens. Consider the derivative of the electron self-energy $\partial\Sigma(p)/\partial p_\mu$. An arbitrary term in the expansion of the electron self-energy has the form

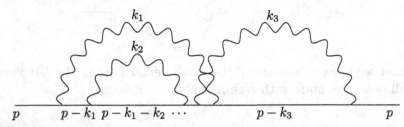

That is, there is one continuous electron line, and the external momentum may be chosen to flow only along this line. Then electron propagators depend on the external momentum via differences $p - k_1$, $p - k_1 - k_2$ etc. The derivative of any electron propagator looks like

$$\frac{\partial}{\partial p_\mu}\frac{1}{m - \hat{p} + \hat{k}} = \frac{1}{m - \hat{p} + \hat{k}}\,\gamma_\mu\,\frac{1}{m - \hat{p} + \hat{k}}.$$

Note that the derivative consists of two propagators and the matrix γ_μ, the latter being nothing but the photon vertex. This means that the differentiation attaches to the self-energy diagram an external vertex of emission of a photon with zero momentum. Graphically, after differentiation we obtain

$$-\frac{\partial\Sigma}{\partial p_\mu} = \text{(diagrams)}$$

or, in the form of the skeleton expansion,

$$-\frac{\partial\Sigma}{\partial p_\mu} = \text{(diagrams)} \qquad (5.18)$$

Thus, we have obtained a linear equation for $\partial\Sigma/\partial p_\mu$. It leads, for example, to the Ward identity (4.56)

$$\gamma_\mu - \frac{\partial\Sigma}{\partial p_\mu} = \Gamma_\mu(p,p,0)$$

which may be easily derived from (5.18) if we simply add the tree vertex to both sides of this equation. Subtraction in (5.18) may be carried out exactly in the same way as for the vertex part. Indeed,

$$\frac{\partial}{\partial p_\mu}G^{-1}(p) = \frac{\partial}{\partial p_\mu}\,[\,m - \hat{p} + \Sigma(p)\,] =$$

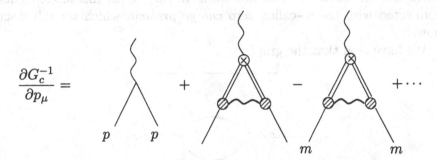

Introducing, as usual, the renormalized Green function G_c according to $G^{-1} = Z_2^{-1}G_c^{-1}$ with $Z_2^{-1} = 1 - \Sigma'(m)$, and adding and subtracting $\Sigma'(m)$ term by term (which is equivalent to the addition and subtraction of $\Lambda(m,m)$), we obtain

This equation contains only the renormalized charge e_c, and its right-hand side is convergent.

For $\partial\Pi(k^2)/\partial k_\mu$ we can construct a similar linear equation and verify that it is also renormalizable.

To summarize, we have shown that all observables can be expressed in terms of the renormalized, physical charges and masses and renormalized Green functions, and we never encounter any divergences. This was proved in perturbation theory, but, strictly speaking, we cannot be sure that it will remain valid outside the perturbative framework.

5.2 The zero charge problem in quantum electrodynamics

We have constructed quantum electrodynamics in the following way. We started with the Green functions of the electron and the photon

$$= G$$

$$= D_{\mu\nu},$$

and considered the simplest interaction

In the case of the π-meson (a scalar charged particle) we also had to introduce

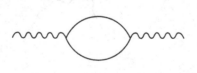

There is no reason to consider more complicated interactions, since the respective theories would be non-renormalizable.

The theory constructed in this way is in excellent agreement with experiment. However, it does not work at very small distances. This is connected with the so-called *zero charge problem* which we will discuss now.

We have seen that the graphs

diverge in the region of large virtual momenta and in this region they are practically independent of the external momenta. For example, the vertex part

$$\Lambda_\mu^{(1)} = e^2 \int \frac{d^4 k}{(2\pi)^4 i} \gamma_\nu \frac{1}{m - \hat{p}_1 + \hat{k}} \gamma_\mu \frac{1}{m - \hat{p}_2 + \hat{k}} \gamma_\nu \frac{1}{k^2} \qquad (5.19)$$

does not depend on p_1 and p_2, when $p_1, p_2 \ll k$. In the region of large k (i.e. small distances) the integral is divergent, and the theory makes no sense. In order to get rid of this problem, we have introduced a large cutoff parameter Λ. The contribution of the large integration momenta k close to Λ does not depend on the external momenta $p_1, p_2 \ll \Lambda$. Next, we considered $\Gamma(m, m, 0)$ and subtracted it from $\Gamma(p_1, p_2, k)$. The result of this subtraction is convergent. We do not know anything about the contribution of the large k region, but we have avoided the problem with the help of the renormalization constants, by hiding our ignorance in the renormalized charge $e_c = Z_1^{-1} Z_2 \sqrt{Z_3} e$.

All these considerations are true, however, only at small external momenta: we have assumed that $p_1, p_2 \ll \Lambda$. What happens if we start to increase the external momenta, i.e. if $p^2/m^2 \gg 1$? (This problem was first raised by Gell-Mann and Low [8] and solved by Landau, Abrikosov and Khalatnikov [9]).

Let us consider the photon polarization operator

$$= \Pi_{\mu\nu}(k) = (g_{\mu\nu} k^2 - k_\mu k_\nu) \Pi(k^2).$$

We have calculated the asymptotic behaviour of the polarization operator in (4.77)

$$\Pi_c^{(1)}(k^2) \simeq \frac{\alpha_c}{3\pi} \ln \frac{k^2}{m^2}.$$

Hence, the first terms of the perturbation theory expansion for the photon Green function at large momenta have the form

$$= \frac{1}{k^2} + \frac{1}{k^2} \left(k^2 \frac{\alpha_c}{3\pi} \ln \frac{k^2}{m^2} \right) \frac{1}{k^2} \sim \frac{1}{k^2} \left(1 + \frac{\alpha_c}{3\pi} \ln \frac{k^2}{m^2} \right),$$

i.e. for large k^2 the series diverges and the perturbation theory does not work. Obviously, the term contains $\ln^2 (k^2/m^2)$ and, generally speaking, more complicated graphs will grow even stronger with the growth of k^2. (The main contribution to the vertex part in (5.19) comes exactly from photons with large virtual momenta $k \sim \Lambda$.)

Let us consider the integral (5.19) in more detail. It contains factors of the type

$$\frac{1}{m - \hat{p}_1 + \hat{k}} \times \cdots$$

The number of denominators increases as the diagrams become more complicated, and in the same way the power of the external momenta in the denominator grows. The integrand decreases with the growth of p, and the integrands for more complicated diagrams decrease faster. This decrease will be significant, however, only for very large external momenta $p \sim \Lambda$, since the main contribution to the integral comes from $k \sim \Lambda$.

Consider the intermediate region $m^2 \ll |p^2| \ll \Lambda^2$ for sufficiently large Λ. In this case (5.19) has a simple form

$$\Lambda_\mu^{(1)} \sim e^2 \int_p^\Lambda \frac{d^4k}{k^4} \sim \alpha_0 \ln \frac{\Lambda^2}{p^2},$$

since the main contribution to the integral (5.19) comes from the region $p \ll k \ll \Lambda$. Let us choose Λ so that

$$\alpha_0 \ln \frac{\Lambda^2}{p^2} \sim 1, \quad \text{in spite of } \alpha_0 \ll 1. \tag{5.20}$$

The vertex then has the following structure:

$$\Gamma = \sum_{n \geq m} c_{nm} \, \alpha_0^n \ln^m \frac{\Lambda^2}{p^2}$$

$$= 1 + \alpha_0 \ln \frac{\Lambda^2}{p^2} + \alpha_0^2 \ln^2 \frac{\Lambda^2}{p^2} + \alpha_0^2 \ln \frac{\Lambda^2}{p^2} + \cdots \tag{5.21}$$

Obviously, terms of the type $\alpha_0^n \ln^n \frac{\Lambda^2}{p^2} \sim 1$ give the largest contribution, while terms of the type $\alpha_0^{n+1} \ln^n \Lambda^2/p^2 \sim \alpha_0$ play the rôle of small corrections. Thus, we can write the expression for the vertex in the form

$$\Gamma = f_1 \left(\alpha_0 \ln \frac{\Lambda^2}{p^2} \right) + \alpha_0 f_2 \left(\alpha_0 \ln \frac{\Lambda^2}{p^2} \right) + \alpha_0^2 f_3 + \cdots \tag{5.22}$$

To simplify the problem, we consider only the first term (this is called the leading logarithmic approximation). Let us determine Γ, G and D in this approximation.

The total vertex part is then just the sum of the first skeleton diagrams (5.13):

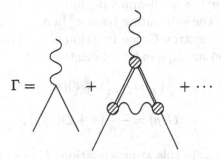

The first simplification is due to the Ward identity $Z_1 = Z_2$ (see (4.51)) which makes the divergences connected with the vertex part and with the electron Green function cancel in the expression for the renormalized charge

$$e_c^2 = Z_1^{-2} Z_2^2 Z_3 \, e_0^2.$$

This means that we should be able to reformulate the theory in such a way that these divergences do not arise at all. This may be achieved by a proper choice of gauge (Landau [10]). In the Landau gauge the photon Green function is

$$D_{\mu\nu}^t = \frac{1}{k^2} \left(g_{\mu\nu} - \frac{k_\mu k_\nu}{k^2} \right). \tag{5.23}$$

Let us show that Γ_μ in this gauge is ultraviolet finite. Consider

$$
\begin{aligned}
\Gamma_\mu^{(1)} &= e_0^2 \int \frac{d^4k}{(2\pi)^4 i} \gamma_\alpha \frac{1}{m - \hat{p}_1 + \hat{k}} \gamma_\mu \frac{1}{m - \hat{p}_2 + \hat{k}} \gamma_\beta D_{\alpha\beta}^t(k) \\
&= e_0^2 \int_p^\Lambda \frac{d^4k}{(2\pi)^4 i} \left[\gamma_\alpha \frac{1}{\hat{k}} \gamma_\mu \frac{1}{\hat{k}} \gamma_\alpha \frac{1}{\hat{k}} - \hat{k} \frac{1}{\hat{k}} \gamma_\mu \frac{1}{\hat{k}} \hat{k} \frac{1}{k^4} \right] \\
&= e_0^2 \int_p^\Lambda \frac{d^4k}{(2\pi)^4 i} \left[\frac{\gamma_\alpha \hat{k} \gamma_\mu \hat{k} \gamma_\alpha}{k^6} - \frac{\gamma_\mu}{k^4} \right].
\end{aligned}
$$

The first term in the square brackets in the integrand contains the factor $k_i k_j$, and due to rotational symmetry the respective integral is proportional to the unit tensor (Kronecker symbol in Euclidean space). Then we can substitute in the integrand

$$\frac{k_i k_j}{k^6} \implies \frac{g_{ij} k^2}{4} \frac{1}{k^6}$$

and use

$$\gamma_\alpha \gamma_i \gamma_\mu \gamma_i \gamma_\alpha = 4 \gamma_\mu.$$

We see that the main contributions to the integral cancel. It is straight-forward to show that in the next order of perturbation theory the leading term $\propto \left(\alpha_0 \ln \lambda^2/p^2\right)^2$ also disappears. In the leading logarithmic approximation we ignore the subleading term $\alpha_0^2 \ln \Lambda^2/p^2$. Similar consideration applies also to the electron Green function G, and in the leading logarithmic approximation at large external electron momentum we obtain

$$\Gamma_\mu = \gamma_\mu \left(1 + \mathcal{O}(e_0^2)\right),$$
$$G(p) = -\frac{1}{\hat{p}}\left(1 + \mathcal{O}(e_0^2)\right). \tag{5.24}$$

Now we turn to $\Pi_{\mu\nu}$ in this approximation. Due to (5.24) the expression for $\Pi_{\mu\nu}$ simplifies:

$$\tag{5.25}$$

We can check by direct calculation of higher order graphs

that (5.25) is valid in the leading logarithmic approximation. The leading contributions induced by these graphs cancel each other. Hence, in the leading logarithmic approximation, calculation of the polarization operator $\Pi_{\mu\nu}$ reduces to calculation of the simplest diagram with bare vertices and electron Green functions.

Let us now derive the subtracted polarization operator $\Pi_{\mu\nu}(k) - \Pi_{\mu\nu}(0)$:

$$\Pi_{\mu\nu}(k) - \Pi_{\mu\nu}(0) = -e_0^2 \int \frac{d^4p}{(2\pi)^4 i} \operatorname{Tr}\left\{\gamma_\mu \frac{1}{m-\hat{p}}\gamma_\nu \left[\frac{1}{m-\hat{p}+\hat{k}} - \frac{1}{m-\hat{p}}\right]\right\}.$$

Expand $1/m - \hat{p} + \hat{k}$ in powers of $\hat{k}$:

$$\frac{1}{m-\hat{p}+\hat{k}} = \frac{1}{m-\hat{p}}$$
$$-\frac{1}{m-\hat{p}}\hat{k}\frac{1}{m-\hat{p}} + \frac{1}{m-\hat{p}}\hat{k}\frac{1}{m-\hat{p}}\hat{k}\frac{1}{m-\hat{p}} + \cdots \tag{5.26}$$

The first term in (5.26) cancels in the square brackets in the integrand. The second term vanishes after integration due to rotational symmetry:

$$\int \frac{d^4p}{(2\pi)^4 i} \operatorname{Tr}\left\{\gamma_\mu \frac{1}{\hat{p}}\gamma_\nu \frac{1}{\hat{p}}\hat{k}\frac{1}{\hat{p}}\right\} = \int \frac{d^4p}{(2\pi)^4 i} \operatorname{Tr}\frac{\gamma_\mu \hat{p}\gamma_\nu \hat{p}\hat{k}\hat{p}}{p^6} = 0.$$

The third term in (5.26) gives just the logarithmic divergence, and we obtain

$$\Pi_{\mu\nu}(k) - \Pi_{\mu\nu}(0) = - e_0^2 \int \frac{d^4 p}{(2\pi)^4 i} \, \mathrm{Tr}\left\{ \gamma_\mu \frac{1}{\hat{p}} \gamma_\nu \frac{1}{\hat{p}} \hat{k} \frac{1}{\hat{p}} \hat{k} \frac{1}{\hat{p}} \right\}. \qquad (5.27)$$

Due to the transverse structure of the polarization operator, it depends only on one scalar function

$$\Pi_{\mu\nu} = (g_{\mu\nu} k^2 - k_\mu k_\nu) \Pi(k^2), \qquad \Pi_{\mu\mu} = 3k^2 \Pi(k^2).$$

On the other hand, it is easy to calculate the trace of the integrand in (5.27):

$$\gamma_\mu \frac{1}{\hat{p}} \gamma_\mu \frac{1}{\hat{p}} = \frac{\gamma_\mu \hat{p} \gamma_\mu \hat{p}}{p^4} = -\frac{2p^2}{p^4} = -\frac{2}{p^2},$$

and

$$3k^2 \Pi(k^2) = 2e_0^2 \int \frac{d^4 p}{(2\pi)^4 i} \cdot \frac{1}{p^2} \, \mathrm{Tr}\left\{ \hat{k} \frac{1}{\hat{p}} \hat{k} \frac{1}{\hat{p}} \right\} = 2e_0^2 \int \frac{d^4 p}{(2\pi)^4 i} \frac{\mathrm{Tr}\{\hat{p}\,\hat{k}\,\hat{p}\,\hat{k}\}}{p^6}.$$

After simple transformations, we derive

$$\Pi(k^2) = -\frac{4}{3} e_0^2 \int \frac{d^4 p}{(2\pi)^4 i} \frac{1}{p^4},$$

where the integration momentum is larger than the momentum of the external photon. Rotating the contour of integration $ip_0' = p_0$, we arrive at an integral over the four-dimensional Euclidean space

$$\Pi(k^2) = -\frac{4}{3} e_0^2 \int_k^\Lambda \frac{d^4 p}{(2\pi)^4} \frac{1}{p^4}. \qquad (5.28)$$

In spherical coordinates

$$d^4 p = p^2 dp^2 \frac{d\Omega}{2} = \pi^2 p^2 dp^2,$$

and we finally obtain

$$\Pi(k^2) = -\frac{4e_0^2}{3 \cdot 16\pi^2} \ln \frac{\Lambda^2}{|k^2|} = -\frac{\alpha_0}{3\pi} \ln \frac{\Lambda^2}{|k^2|}. \qquad (5.29)$$

Thus, in the leading logarithmic approximation the unrenormalized photon Green function has the form

$$D_{\mu\nu} = \frac{g_{\mu\nu}}{k^2} \frac{1}{1 - \Pi(k^2)} = \frac{g_{\mu\nu}}{k^2} \frac{1}{1 + \frac{\alpha_0}{3\pi} \ln \frac{\Lambda^2}{|k^2|}} \equiv \frac{g_{\mu\nu}}{k^2} d. \qquad (5.30)$$

Let us renormalize this expression. First we write for the scalar function d

$$d^{-1} = 1 + \frac{\alpha_0}{3\pi} \ln \frac{\Lambda^2}{|k^2|} = 1 + \frac{\alpha_0}{3\pi} \ln \frac{\Lambda^2}{m^2} - \frac{\alpha_0}{3\pi} \ln \frac{|k^2|}{m^2}$$

$$= \left(1 + \frac{\alpha_0}{3\pi} \ln \frac{\Lambda^2}{m^2} \right) \left(1 - \frac{\frac{\alpha_0}{3\pi} \ln \frac{|k^2|}{m^2}}{1 + \frac{\alpha_0}{3\pi} \ln \frac{\Lambda^2}{m^2}} \right), \tag{5.31}$$

and introduce the renormalization factor Z_3

$$Z_3^{-1} = 1 + \frac{\alpha_0}{3\pi} \ln \frac{\Lambda^2}{m^2}. \tag{5.32}$$

Then the function d becomes

$$d = Z_3 \frac{1}{1 - \frac{Z_3 \alpha_0}{3\pi} \ln \frac{-k^2}{m^2}} = \frac{Z_3}{1 - \frac{\alpha_c}{3\pi} \ln \frac{-k^2}{m^2}}. \tag{5.33}$$

We see that the cutoff parameter is now embodied into the overall renormalization constant Z_3 while in the denominator it is swallowed by the physical charge e_c. The very fact that we have succeeded in eliminating the cutoff momentum Λ and arrived at the multiplicative dependence on Z_3 reflects the renormalizability of electrodynamics.

We also obtained a nice relationship between the renormalized and bare charges:

$$\alpha_c = \frac{\alpha_0}{1 + \frac{\alpha_0}{3\pi} \ln \frac{\Lambda^2}{m^2}}. \tag{5.34}$$

At first glance, this result looks reasonable: the renormalized charge α_c is less than the bare charge α_0, $\alpha_c < \alpha_0$, as it should be due to vacuum polarization. However, if we go to the limit $\Lambda \to \infty$, we get

$$\alpha_c \simeq \frac{3\pi}{\ln \frac{\Lambda^2}{m^2}} \to 0, \qquad \text{with} \quad \Lambda \to \infty, \tag{5.35}$$

i.e. any bare charge α_0 is screened completely (recall that we considered only the case $\alpha_0 \ll 0$) if it is shrunk to a point. In other words, the physical charge is always zero, $\alpha_c = 0$. This could mean that our approach is wrong at short distances. On the other hand, if there exists such small scale where QED is not valid any more, we can calculate α_c in terms of this scale and, vice versa, we can determine Λ from the value of α_c (since we know $\alpha_c \simeq 1/137$). From (5.35) we obtain

$$\frac{\Lambda^2}{m^2} \simeq \exp \left\{ \frac{3\pi}{\alpha_c} \right\},$$

or, numerically, $1/\Lambda \simeq 10^{-50}$ cm.

The concrete value of this small scale changes somewhat if one includes contributions of different sorts of particles to the vacuum polarization. If, for instance, there are ν species of charged spin $\frac{1}{2}$ particles, then we have instead of (5.34)

$$\alpha_c = \frac{\alpha_0}{1 + \nu \frac{\alpha_0}{3\pi} \ln \frac{\Lambda^2}{m^2}}, \qquad (5.36)$$

and the value of Λ changes correspondingly.

Assuming that QED is not valid at a scale of the order of the Planck length,* i.e. at $\ell_P \simeq 10^{-33}$ cm, we come to the conclusion that the number of possible sorts of charged elementary particles is

$$\nu \simeq 12.$$

The theoretical situation for the photon Green function looks even worse than the problem of zero physical charge. For any value of the physical charge α_c the photon Green function

$$d_c = \frac{1}{1 - \frac{\alpha_c}{3\pi} \ln \frac{-k^2}{m^2}}$$

acquires a pole at some large space-like momentum $k^2 < 0$, and this implies the existence of a particle with an imaginary mass. In a sense, this is an artificial problem, since in our theory $\Lambda = \infty$, and hence $\alpha_c = 0$. This means that actually there is no unphysical pole, but then there is no interaction either!

This problem is not yet solved.[†]

---

* The Planck length is $\ell_P = \sqrt{G}$, where G is the Newton gravitational constant.

† V. N. Gribov left a draft of 'QED at short distances' which he was preparing as additional sections for this chapter. He was planning to discuss a possible solution of the Landau pole–zero charge problem in quantum electrodynamics or, taken more widely, in the Glashow–Weinberg–Salam theory which unifies electrodynamics and weak interactions. This solution came, if one may say so, as a by-product of his 20-year study of the problem of quark confinement in quantum chromodynamics (QCD) – the microscopic theory of 'coloured' quarks and gluons believed to be responsible for the structure of hadrons and their interactions. Gribov found [11] that when the coupling exceeds a critical value,

$$\frac{\alpha}{\pi} > 1 - \sqrt{\frac{2}{3}},$$

the theory changes drastically. The so-called supercritical binding of fermions takes place which leads to the appearance of bound states with negative total energy, so that the perturbative vacuum becomes unstable. A phenomenon similar to a phase transition in solid state physics occurs, and the dynamics of the theory becomes essentially different.

In QCD the colour coupling between quarks and gluons, contrary to $\alpha_{\mathrm{e.m.}}$, *increases* with distance and hits the critical value at 'large distances' of about 1 fermi =

Due to the smallness of the coupling constant α_c, in quantum electrodynamics this difficulty arises at academically small distances, and is irrelevant for real physical problems. In non-asymptotically free theories of strong interactions we have $g \sim 1$ and face this problem immediately, as we are forced to introduce an ultraviolet cutoff parameter $\Lambda \sim m$. Here the problem becomes real and severe.

Let us note that to arrive at the zero charge result, we have used the logarithmic approximation

$$ e_0^2 \ln \frac{\Lambda^2}{p^2} \sim 1 \,, \qquad e_0^2 \ll 1. $$

Strictly speaking, this means that within our approximation we are not allowed to take the limit $\Lambda \to \infty$. However, according to Pomeranchuk [14] only the renormalized charge α_c (5.34) enters higher order unaccounted for corrections. Since $\alpha_c \to 0$ at $\Lambda \to \infty$, all corrections also vanish in this limit. This means that our conclusion about the interaction vanishing in quantum electrodynamics does not depend on the condition $e_0^2 \ln \left(\frac{\Lambda^2}{p^2} \right) \sim 1$. The only necessary hypothesis is

$$ e_0^2 \ll 1. $$

Abandoning the latter condition would mean, however, that from the very beginning there was no perturbation theory and, therefore, quantum electrodynamics has not been formulated.

10^{-13}cm. Gribov argued that the supercritical binding of light quarks results in the instability of colour states, that is, in the confinement of colour (see [12]).

In the context of quantum electrodynamics, the supercritical binding phenomenon develops at extremely short distances, of the order of the Planck scale.

On the one hand, it may be responsible for the appearance of the Higgs scalar boson, much wanted for the consistency of the electroweak GWS theory. Within this picture, Gribov predicted the mass of the composite Higgs boson to be slightly larger than that of the heaviest ('top') quark, $m_H \simeq 200\, m_{\text{proton}}$ [13].

On the other hand, the formal Landau pole problem in QED has been resolved: the coupling increases but remains finite at arbitrarily small distances.

'QED at short distances' will be included in a collection of Gribov's works on gauge theories 'Gauge Theories and Quark Confinement', Phasis Publishing house, Moscow, 2001 [www.aha.ru/~phasis].

References

[1] L. D. Landau, E. M. Lifshitz, Quantum Mechanics, Vol. 3, Butterworth–Heinemann, 1997

[2] E. Fermi, Z. Phys. **29**, 315 (1924); C. Weizsäcker, Z. Phys. **88**, 612 (1934); E. Williams, Phys. Rev. **45**, 729 (1935); Dan Vid. Selck. **13**

[3] J. Schwinger, Phys. Rev. **73**, 416 (1948)

[4] R. P. Feynman, Phys. Rev. **74**, 1430 (1948)

[5] Ya. B. Zeldovich and V. S. Popov, Usp. Fiz. Nauk **105**, 403 (1971); Sov. Phys. Usp. **14**, 673 (1972)

[6] A. B. Migdal, Usp. Fiz. Nauk **123**, 369 (1977); Sov. Phys. Usp. **20** 879 (1977)

[7] W. E. Lamb and R. C. Retherford, Phys. Rev. **72**, 241 (1947)

[8] M. Gell-Mann, F. Low, Phys. Rev. **95**, 1300 (1954)

[9] L. D. Landau, A. A. Abrikosov, I. M. Khalatnikov, Dokl. Akad. Nauk USSR **95**, 773 (1954) [in Russian]; Collected papers of L. D. Landau, ed. D. Ter Haar, Gordon and Breach, NY (1965) p. 616

[10] L. D. Landau in Niels Bohr and Development of Physics, ed. W. Pauli, Pergamon Press, London, (1955) p. 52 ; Collected Papers of L. D. Landau, ed. D. Ter Haar, Gordon and Breach, NY (1965) p. 634; L. D. Landau and I. Ya. Pomeranchuk, Dokl. Akad. Nauk USSR **102**, 489 (1955)

[11] V. N. Gribov, Lund preprint LU-TP 91-7 (1991)

[12] V. N. Gribov, in Proceedings of the International School of Subnuclear Physics, 34th course, Erice, Italy (1996); Eur. Phys. J. C **10**, 71 (1999), hep-ph/9807224; Eur. Phys. J. C **10**, 91 (1999), hep-ph/9902279

[13] V. N. Gribov, Phys. Lett. B **336**, 243 (1995)

[14] I. Ya. Pomeranchuk, Dokl. Akad. Nauk USSR **103**, 1005 (1955)

Index